Advances in Entomology

Edited by Ramón Eduardo Rebolledo Ranz

Published in London, United Kingdom

Advances in Entomology
http://dx.doi.org/10.5772/intechopen.1005948
Edited by Ramón Eduardo Rebolledo Ranz

Contributors
Alan Lee Knight, Ashok Kumar Chaubey, Chastity Perry, Dingze Mang, Esteban Basoalto, Hafza Wajeeha Ijaz, Heena Khan, Ijaz Ali, Inam Ullah, Michele Preti, Minglin Lang, Naeem Ullah, Nitesh Kumar, Rafiur Rahman, Sriyanka Lahiri, Yudie Ma

First published in London, United Kingdom, 2025 by IntechOpen
IntechOpen is the global imprint of INTECHOPEN LIMITED, registered in England and Wales, registration number: 11086078, 167-169 Great Portland Street, London, W1W 5PF, United Kingdom

For EU product safety concerns: IN TECH d.o.o., Prolaz Marije Krucifikse Kozulić 3, 51000 Rijeka, Croatia, info@intechopen.com or visit our website at intechopen.com.

British Library Cataloguing-in-Publication Data
A catalogue record for this book is available from the British Library

Advances in Entomology
Edited by Ramón Eduardo Rebolledo Ranz
p. cm.
Print ISBN 978-1-83634-470-4
Online ISBN 978-1-83634-469-8
eBook (PDF) ISBN 978-1-83634-471-1

If disposing of this product, please recycle the paper responsibly.

Meet the editor

Dr. Ramón Eduardo Rebolledo Ranz is an Agronomist Engineer (Universidad Austral de Chile, 1986) and holds a Ph.D. from the Universidad Politécnica de Madrid. A professor since 1986, he teaches Agricultural Entomology, Apiculture, and Environmental Entomology at undergraduate, master's, and Ph.D. levels. He has supervised over 100 theses (undergraduate and postgraduate) and participated in numerous research projects. Dr. Rebolledo has published around 100 scientific articles in international journals, authored several books and chapters, and edited 10 books. He has advised the Chilean Beekeeping Network, contributed to beekeeping and entomology legislation, and consulted for agricultural companies. He also serves as a reviewer for international journals, having evaluated over 200 papers in his field of expertise.

Contents

Preface **IX**

Chapter 1 **1**
Advances in Vector Control: What Is Still Missing?
by Rafiur Rahman, Naeem Ullah and Ijaz Ali

Chapter 2 **19**
Physiological Roles of Chemosensory Receptors in Lepidoptera
by Dingze Mang

Chapter 3 **47**
Sprayable Sex Pheromones for Codling Moth: Waiting for Grower Adoption
by Alan Lee Knight, Esteban Basoalto and Michele Preti

Chapter 4 **69**
Novel IPM Tools for Twenty-First Century Tree Fruits
by Alan Lee Knight, Michele Preti and Esteban Basoalto

Chapter 5 **91**
Classification of Insect Parasitic Nematodes and Potential of Entomopathogenic Nematodes as Biocontrol Agents against Insect Pests in Agriculture
by Heena Khan, Nitesh Kumar and Ashok Kumar Chaubey

Chapter 6 **117**
Expanding Host Range of Invasive Pest *Scirtothrips dorsalis* Hood in the U.S.A.
by Sriyanka Lahiri and Chastity Perry

Chapter 7 **129**
The Role of Metals in Biological Control of Insects
by Inam Ullah, Hafza Wajeeha Ijaz, Yudie Ma and Minglin Lang

Preface

It is a great pleasure to present this new book on insects and the latest research on fundamental aspects, including insect control, pests, and pest studies.

In this text, the reader will find answers to their curiosities and citations of work on research projects and undergraduate and postgraduate theses.

In the book, the chapter authors present the results of their research, as well as updated literature reviews on insects, such as advances in insect vector control, classification of entomopathogenic nematodes, broadening the range of pests in the U.S.A., the role of metals in insect control, tools for integrated pest management (IPM), and the role of chemoreceptors in insects and pheromones in moth pest management. All of these topics are addressed at the highest scientific level by researchers from around the world.

This book can be used and cited by undergraduate students in their entomology and zoology courses, as well as by postgraduate students, university lecturers, and researchers in universities, research institutes, governmental institutions, and the general public.

For the editor, it was a great honour to review all the contributions to this volume, especially given the international scope of the authors involved.

Finally, the editor congratulates all the authors on their hard work in writing at the highest scientific level and for agreeing to all the suggestions for improving the presentation of their work.

Ramón Eduardo Rebolledo Ranz
University of La Frontera,
Temuco, Chile

Chapter 1

Advances in Vector Control: What Is Still Missing?

Rafiur Rahman, Naeem Ullah and Ijaz Ali

Abstract

In this chapter, we will try to explore the gaps and limitations in vector control interventions. The spectrum of vector-borne diseases is spreading to Europe and North America. Conventional vector control methods are losing their efficacy due to several factors, like global warming, the development of insecticide resistance, unplanned urbanization, and exponential population increase. Several countries are investing in novel vector control strategies along with conventional methods. Conventional control strategies can be made efficient by proper planning and management, as new strategies will take time to be operational. Paratransgenic methods like *Wolbachia* have shown promising results, along with other population replacement and population suppression techniques. But these methods cannot stop the spread of insect vectors beyond their native range. Among growing concerns hindering vector control, lack of coordination between vector control programs, over-reliance on neurotoxic insecticides, lack of public health education, lack of involvement of important sectors of society, security risks, similar control interventions against different target species, lack of skills of health workers, inadequate duration of a specific control intervention, and lack of political will are important. To effectively control the spread of vector-borne diseases, we should address these issues. Other strategies, like vaccine and drug development along with the One Health approach, are very important. We conclude that conventional control interventions, participation of all sectors of society, effective and prompt communication between stakeholders, and incorporating them into an integrated vector management approach are the only way forward in vector control.

Keywords: dengue, *Wolbachia*, vector control, vector-borne diseases, vaccine

1. Introduction

Vector-borne diseases (VBDs) such as dengue, malaria, chikungunya, and Zika are making headlines in almost every part of the world. Collectively, these diseases threaten every 4 out of 5 individuals, affecting mainly poor populations living in or around the tropics [1]. VBDs range from viral and bacterial to protozoal infections. They include mosquito-borne diseases such as malaria, dengue, chikungunya, Zika, yellow fever, filariasis, snail-borne schistosomiasis, blackfly-borne onchocerciasis, flea-borne plague and Tungiasis, louse-borne typhus and relapsing fever, sandfly-borne leishmaniasis and phlebotomus fever, triatome-borne Chagas disease

(American trypanosomiasis), tsetse-borne sleeping sickness (African trypanosomiasis), tick-borne Crimean Congo hemorrhagic fever (CCHF), borreliosis, spotted fever, Q fever, Tularemia, Lyme disease, etc. According to the UN Office for Disaster Risk Reduction website, these diseases account for more than 17% of all infectious diseases globally, with an annual death toll exceeding 17,000 [2]. In 2022, malaria cases reached an escalating number of 249 million, which is five million more than the previous year, i.e., 2021.

Economically, only malaria results in an annual loss of 12bn USD across Africa [3]. According to the World Health Organization, currently every 8 out of 10 persons is at risk of one or more VBDs [4]. Among vector-borne viral diseases, dengue stands out as the most prevalent flavivirus disease. The African region reported 171,991 dengue cases with 753 deaths only in 2023. Countries more affected included Benin, Burkina Faso, Cabo Verde, Ghana, etc. The actual number of dengue cases in Africa can be higher, as there is no proper diagnostic system in the region for timely detection and/or confirmation, and dengue is similar to several other arboviral diseases. The American region, on the other hand, has a health information platform called Plisa (from Spanish = La Plataforma de infomarción en Salud). According to Plisa, Brazil reported more dengue cases (1359 cases per 100,000 population), followed by Peru (813 cases per 100,000 population) and Mexico (235 cases per 100,000 cases). Severe dengue cases were reported more in Colombia (1.35% of all cases, followed by Brazil (0.05% of total cases), and Mexico (0.54% of all cases). In the eastern Mediterranean region, dengue is present in all countries, including Pakistan, Afghanistan, Iran, India, and Bangladesh.

2. A brief history of vector control

Vector control is the only effective strategy that can be applied at large scales against VBDs [5]. The history of vector control traces back to the Greek and Roman empires (c. 550 B.C.), when large cleaning campaigns were carried out to reduce fever and plague. Mechanical and environmental vector control management, such as using bed nets by Egyptians and Romans, curtains by Indians (Marco Polo's visit to India), and sleeping in tall buildings to avoid mosquito bites, is seen in an era when vectors' involvement in parasite transmission was barely known. In Philadelphia (US), sanitation measures such as cleaning sewers and discarding stored water from ships brought yellow fever cases to a minimum [5].

After Sir Ronald Ross's experiments on malaria, showing the involvement of *Anopheles* mosquitoes in malaria parasite transmission, a lot of work has been done [6]. In the early days after Ross's work, the primary vector control intervention was source elimination, which ultimately reduced the Anopheles population. Housing modifications, such as netting of doors and windows, were advised to interrupt contact between the host and the mosquito.

Source reduction/environmental management reduced malaria cases in present-day Malaysia, where he targeted breeding sites of the local malaria vector (*An. umbrosus*). He also exposed the possible sites to the sun, which eliminated malaria in the region to a great extent. In Singapore, the malaria vector *An. maculatus* was controlled, and malaria cases were brought to zero by the same intervention. This intervention was looked after by a British doctor, Sir Malcolm Watson, who joined MMS (Malayan Medical Services) in 1900 [7]. This intervention, which later got his name "species sanitation" by Dutch zoologist Nicolaas Swellengrebel, who replicated it for Indonesian territories between 1920 and 1935 [7].

In 1901, William Gorgas, along with Joseph Le Prince, started the Ae. aegypti control program, following the discovery of the involvement of this mosquito species in the spread of the yellow fever virus. Vector control interventions in this period included inspection of residential areas and houses, destruction of breeding sites or oiling them, and fumigating their surroundings with sulfur or pyrethrum. This program later on included Anopheles control by using the same measures against Aedes, but including the use of larvivorous fish and restricting animals from feeding in the outfield to stop creating breeding places in flooded footprints. Due to these interventions, mortality due to yellow fever dropped by 96.22% and due to malaria by 99.22% despite a great expansion in human population [8].

An. gambiae was completely eradicated from northern Egypt through a weekly larviciding campaign and house spraying. Larviciding was done with Paris green, and houses were sprayed with pyrethrum. Trains, planes, and other automobiles were treated with residual chemicals to prevent *An. gambiae* from spreading further north, bringing malaria cases fall by 99% from 1942 to 1946 [9].

Notable vector control interventions against malaria and yellow fever vectors can be seen in Havana, led by Joseph White, Oswaldo Cruz in Rio de Janeiro, and Emilio Ribas in the Brazilian city of Santos. All these interventions were inspired by the work of Gorgas, the chief sanitary officer of the Panama Canal, who was supported by Le Prince, where they eliminated yellow fever and brought malaria cases to a history low [8, 10].

Sandfly control was impressed by the fact that a sandfly bite itself is a cause of nuisance in soldiers during the WWI. At the time, leishmaniasis was not yet characterized as the primary disease vectored by sandflies, yet sandfly fever was the main disease associated with phlebotomines. Control measures against sandflies included interventions to make the ground impermeable for sandflies to lay eggs and nourish. British soldiers pitched their camps away from refuse and leveled drained ground, sealed rocks with cresol or sawdust. Chemical control measures included spraying tents and buildings using cresol, crude oil, or paraffin. Normal bed nets were replaced with fine mesh size to stop sandflies' entry [11, 12].

In the post-WWII era, dichloro-diphenyl-trichloroethane (DDT) and pyrethrum were used as first-choice residual insecticides against malaria and other vectors. After the conclusion of the war, insecticides were deployed in many regions of the world, including Europe, to fight typhus, caused by Rickettsia prowazekii and transmitted by body lice [13]. Malaria eradication also became more realistic with the usage of DDT in many regions, including the US, the Caribbean, Asia, Europe, etc. In 1949, the US was declared malaria-free as a significant public health concern [14]. The Ross-Macdonald model, the establishment of the Global Malaria Eradication Program (GMEP), and other environmental management programs followed through. Malaria in Europe was eradicated thanks to GMEP. In 1947, Brazil started its *Ae. aegypti* eradication program. Several control interventions, including inspection of breeding containers, management of larval breeding sites, and residual spraying with DDT. These interventions helped eradicate *Ae. aegypti*, and, consequently, yellow fever from several South American nations, including Bolivia, Brazil, Chile, Costa Rica, Ecuador, Guatemala, Honduras, Panama, Paraguay, Peru, and Uruguay [15] [Political Support]. With the eradication of yellow fever, political support and investment shrank in these countries, despite many warnings from the scientific community [16]. This lack of political will and investment resulted in weak vector surveillance, and hence *Aedes* returned to many countries from where *Ae. aegypti* was not eradicated, like Argentina, the US, Venezuela, etc.

Reduction in phlebotomine populations can be attributed to malaria control programs, especially GMEP in the 1950s; however, specialized vector control approaches took place in China and Brazil (Evandro Chagas initiated the use of DDT against leishmania vectors in Brazil in 1936), decreasing cases of both Visceral (VL) and Cutaneous (CL) leishmania cases down (main article) [17].

After the ban on GMEP, malaria was poorly controlled in many parts of the world. Many outbreaks took place in Asia, Brazil, etc. Economic crisis resulted in an increase in insecticide prices, resistance to insecticides and drugs increased as a result of poor strategies, and operational capacity decreased [18]. A WHO-led intervention employed propoxur and MDA (mass drug administration) decreased malaria cases from 80 to 30% [19]; however, the intervention did not continue, and malaria bounced back in the following year.

The World Health Assembly (WHA) in 1978, 1979, and 1993 endorsed a more technical approach toward malaria elimination. This approach is based on biological, social, ecological, and economic determinants of malaria rather than merely using insecticides and MDA. This strategy called for a selective approach in applying whether, where, and which intervention to apply for vector control.

Synthetic pyrethroids on Insecticides Treated Nets (ITNs) were introduced in vector control, although several communities were already sleeping under bed nets. In 2007, WHO recommended the use of ITNs for the whole community instead of high-risk groups, either free or at a subsidized price. The WHA reiterated that malaria control in Sub Saharan Africa (SSA) can lead to malaria cases globally. IRS was used in a very limited area, mainly funded by the US President's Malaria Initiative [20]. ITNs and IRS brought malaria cases down to 50% [21], of which a 68% decrease is attributed to ITNs and 13% to IRS.

2.1 Arboviruses, a new global threat

In the nineteenth and twentieth centuries, protozoans and nematodes were the major vector-borne parasites; the twenty-first century is marked by the outbreaks of dengue, chikungunya, Zika, yellow fever, etc. All these diseases share a common phenomenon; they are all vectored by *Aedes*. They are sometimes referred to as *Aedes*-borne diseases [22, 23]. The expansion of vector species beyond their native territory is the main reason for the global spread of these diseases. We will discuss in detail the factors that contribute to their worldwide spread.

3. Beyond insecticides, alternative vector control approaches

If we talk about current innovative control interventions, we will use the current WHO Vector Control Advisory Group (VCAG) portfolio. Methods with promising results include autodissemination of pyriproxyfen, pyrethroid-treated nets, attractive targeted sugar baits or ATSBs, genetic manipulation of mosquitoes, the World Mosquito Program's *Wolbachia* method, spatial repellents, interventions under the umbrella of Sterile Insect Technique (SIT), insecticides treated eave tubes, gene drive (defined by VGAG as genetic manipulation of mosquitoes through gene drive technology to reduce vector populations and transmission), etc. [24].

So far, vector control strongly relies on the use of neurotoxic insecticides, mainly of the pyrethroid class. These insecticides pose a constant selection pressure, which helps to select resistant populations [25]. As mosquitoes go through complete metamorphosis, only larval and adult stages can be controlled. But larval control has been the most effective tool compared to adult control [26]. Spinosad, derived from *Saccharopolyspora spinosa,* which is a soil actinomycete, has two insecticidal agents, A and D. Both these agents are effective against all types of mosquito larvae tested so far [27]. Other entomopathogenic fungi, *Beauveria bassiana* and *Metarhizium anisopliae,* have shown promising results against both larvae and adults of *Ae. aegypti* [28].

Bacteria like *Bacillus thuringiensis* var. *israelensis* and *Lysinibacillus sphaericus* [29] have shown promising results. These bacteria release toxins that, upon entering the larvae of mosquitoes, disrupt cell membranes there. These bacteria are effective against *Aedes*, *Culex,* and *Anopheles* mosquitoes, with mortality varying between different genera and species [28].

Natural predators like copepods (*Macrocyclops albidus*), larvivorous fishes, and *Toxorhynchites* larvae, etc., having good output in the field, fall into the category of biological control. Experiments carried out in different parts of America, Asia, and Australia have shown that a single copepod can kill more than forty Aedes larvae per day. In New Orleans, Vietnam, Florida, etc., copepods have shown promising results, particularly against container mosquitoes [30].

Insect growth regulator pyriproxyfen has also shown very promising results against *Ae. aegypti*. In a pilot study in Peru and Italy, pyriproxyfen is distributed to the larval breeding sites through contaminated males, a method called autodissemination. This intervention has significantly reduced *Ae. aegypti* and *Ae. albopictus* populations [31, 32]. IGRs like pyriproxyfen, methoprene, and diflubenzuron, etc., do not directly kill insects but rather delay their growth, hence exposing them more to environmental stress and predators. They do so by mimicking the true growth hormones, changing their growth, development, physiology, and behavior [33].

3.1 *Wolbachia* as an excellent example of a recent vector control intervention

Based on operational applications, *Wolbachia* can be used for both population suppression and population replacement. *Wolbachia* is capable of disturbing vectors' reproduction or reducing their ability to transmit parasites of public health significance.

The *Wolbachia* method is the best example of an innovative vector control intervention.

3.1.1 Wolbachia used for vector population replacement/substitution

Transinfecting *Ae. aegypti* with *Wolbachia* bacteria is an example of population replacement or substitution. The wild population of *Ae. aegypti* is replaced with laboratory-reared *Ae. aegypti,* which are modified (transinfected) with *Wolbachia*. *Wolbachia* bacteria compete for resources with several parasites like dengue virus, chikungunya virus, and *Plasmodium*, hence making vector species less likely to transmit them [34]. A parasite must replicate (in case of viruses) or complete part of its life cycle (as in the case of *Plasmodium*) within the arthropod vector, for which they need resources. When a mosquito carries *Wolbachia*, it creates an environment within the insect that is less hospitable to the growth of parasites. This competition limits the virus's or other parasites' ability to replicate, ultimately reducing the mosquito's competence to transmit these parasites [35, 36].

3.1.2 Wolbachia as a vector population suppression tool

Wolbachia interferes with normal physiological and developmental functions, reducing the survival of vectors. This phenomenon is particularly true and tested for *Ae. aegypti* [37]. This intracellular bacterium also interferes with the normal blood-feeding of *Ae. aegypti* females, hence reducing the chances of egg-laying [38]. *Wolbachia*'s most commonly induced vector population suppression tool so far is cytoplasmic incompatibility (CI). When a *Wolbachia*-infected male crosses with a wild, uninfected female, the resulting progeny is sterile and is called unidirectional CI. In a bidirectional CI, all crosses between *Wolbachia*-infected individuals lead to sterile progeny, provided mosquitoes are infected by different strains. All other crosses are fertile.

Figure 1 shows different types of crosses and their outcomes.

As *Wolbachia* infection is causing sterility, it can be referred to as Sterile Insect Technique (SIT), but if the *Wolbachia* strain is not established in the target vector population, with the males as a dead-end host, we refer to this technique as IIT or Incompatible Insect Technique. *Wolbachia* strains established so far include *w*Pip and *w*Mel isolated from *Culex* and *Drosophila*, respectively [39, 40]. According to the World Mosquito Program's official website, several countries, including Australia, Brazil, Colombia, Singapore, etc., have achieved significant results using a unidirectional approach, releasing Wolbachia-infected male mosquitoes. In Brazil, for example, lab-infected, *Wolbachia*-carrying mosquitoes are called "WOLBITOS" instead of mosquitoes. In Australia, males infected with the *w*Mel strain and released into the field; after a few months of release, *w*Mel has established itself in *Ae. aegypti* populations [34, 39, 41].

SIT on its own is a method in which a large number of sterile males are released in the field. Male sterility is achieved by exposing them to certain chemicals or gamma radiation. Both of them cause permanent damage to an insect's chromosome or dominant lethal mutation in sperm. SIT has been extremely effective against *An. albimanus* and *Cx. quinquefasciatus* in El Salvador and Florida, respectively [42].

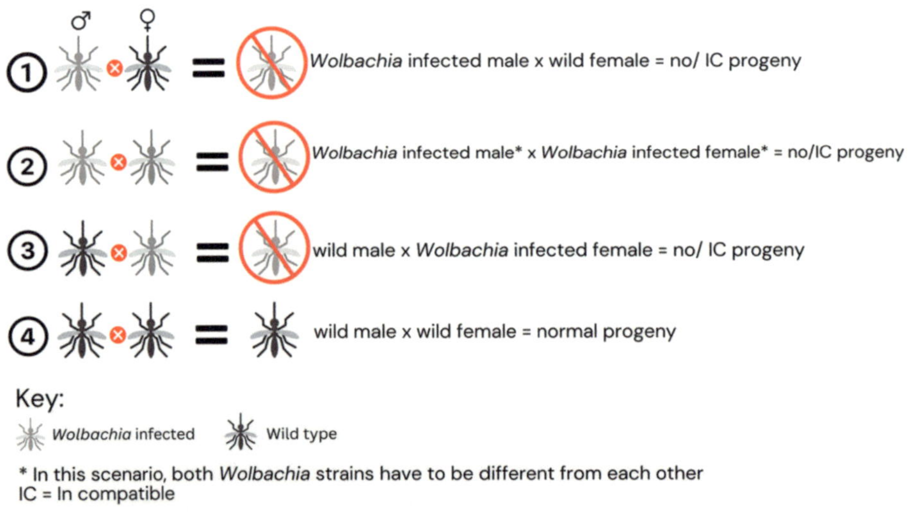

Figure 1.
Different types of crosses possible between Wolbachia*-infected male and female mosquitoes.*

As sterilizing males through the above-mentioned methods can be challenging, a new genetic method, based on RNA interference, aimed at silencing the testis genes and a female sex determination gene with the help of a dsRNA. The dsRNA is provided to larvae in their feed as dsRNA-baited larval food [43, 44].

Other methods like Release of insects carrying a lethal dominant gene (RIDL), RNAi, and homing endonuclease gene (HEG) also fall under the category of genetic control. RIDL basically is a type of SIT, with the principle of female killing. In this method, males carry and deliver female-acting transgenes to wild populations, which may induce mortality in immature stages (pupae) or mature adults. They may also reduce a gene responsible for wing muscle development, disturbing normal flight, resulting in abnormal feeding and other physiological functions [37]. (ISSUE = Large number of males). This method has been in the field in the Cayman Islands [45] and Brazil (OX513A) [46], with promising results.

The RNAi method, instead of killing the mosquito vector, improves the insect's immune response, degrading viral RNA. *Ae. aegypti* have been genetically modified by constructing an effector gene that targets and disrupts DENV2. This is achieved by the expression of a dsRNA that is complementary to the DENV2 genome. The DENV-specific dsRNA triggers the RNAi effect and blocks viral replication inside mosquitoes, making female Aedes mosquitoes resistant to the dengue virus [37].

In HEGs method, specific DNA sequences are recognized and cut. As the name suggests, homing endonuclease genes, or HEGs, encode endonuclease enzymes that, in a heterozygote individual, cut themselves and insert into the exact cop, converting an HEG heterozygote to an HEG homozygote, resulting in an increase in HEG copies in mosquitoes. HEGs can be designed to target different genes like sex-determining, vector competence, and/or fertility [22].

Scientists are in search of delivering insecticides to target species through different routes. The use of Attractive Toxic Sugar Baits (ATSB) is one such method that exploits insects' habit of sugar/nectar feeding [47]. In such baits, a solution is used that consists of a neurotoxic insecticide (Spinosad) or an oral toxin (boric acid) [48] or even an IGR (pyriproxyfen) [49]. ATSBs have shown good results in field and semi-field conditions when used against *Ae. aegypti* and *Ae. albopictus* [50].

3.2 Artificial intelligence (AI) and machine learning (ML) in VBDs

The field of VBDs has seen a surge of interest in AI and ML techniques. While these technologies have shown some promising progress in improving surveillance and modeling, their direct application to control vectors is still very limited [51–53].

4. What is still missing, why are we not able to control VBDs?

Despite many advances in vector control, we are still not able to stop the spread of VBDs. There are several limitations to each vector control method, and we are going to discuss each of them briefly. We have summarized them in **Table 1** as well.

4.1 Insecticide resistance

We cannot deny the fact that the only option in case of an emergency caused by VBDs is neurotoxic insecticides. However, over-reliance on them has led to the development of insecticide resistance [25, 54]. Insecticide resistance, as defined by the

Category	Issues	Solutions
Operational limitations	Community engagement and health education	Conducting awareness sessions in community centers, schools, cultural festivals, etc.
	Lack of political will	Advocate for policy changes and integrate vector control into health agendas.
	Absence of effective vector control surveillance	Establish real-time surveillance systems with modern technology.
Technical limitations	Insecticide resistance	Develop and rotate new insecticides or implement alternative control methods.
	Ignoring vector behavior	Incorporate behavioral studies into control program designs.
	Presence of more than one vector	Use strategies targeting multiple vectors.
	Lack of an integrated approach in vector control	Integrated Vector Management (IVM).
	Sustainability and cost-effectiveness of novel interventions	Invest in scalable and affordable solutions, such as community-based programs.
Logistical limitations	Sustainability and cost-effectiveness of novel interventions	Establish partnerships for funding and resource sharing across regions.
Environmental limitations	Climate change	Develop adaptive vector control measures based on climate modeling.
	Unplanned urbanization	Promote urban planning policies that minimize vector breeding habitats.

Table 1.
Limitations in vector control and their possible solutions.

WHO, is "the ability in a strain of some organism to tolerate doses of a toxicant which would prove lethal to majority of individuals in a normal population of the same species" [55]. A. W. A. Brown defines insecticide resistance as "any population, within a species normally susceptible to a given insecticide, that is no longer controlled by the pesticide in the area concerned" [56]. Mosquito populations worldwide have evolved a diverse array of mechanisms to cope with neurotoxic insecticides. These mechanisms range from behavioral changes to cuticular modifications to detoxification of insecticide molecules to target site modifications [57]. A protein present in the legs of *An. gambiae* was found to be overexpressed in the legs of pyrethroid-resistant *An. gambiae* [58]. This type of resistance is common in regions with extended use of bed nets impregnated with pyrethroids, like Cameroon, Burkina Faso, and Guinea. This shows how smart these mosquitoes are.

Mutations associated with insecticide resistance are reported all over the world; some of them are universal, while others are region-specific. Insecticides like pyrethroids and organochlorines targeting voltage-gated sodium channels (*vgsc* or Na_v) select for *kdr* mutations, which are common in dengue and malaria vectors. Among them, F1534C is found in *Ae. aegypti* populations resistant to pyrethroids. Other mutations like 410 L, 989S, 1016I/G, and 1520I are region-specific [59]. The origin of the 1534 mutation, when studied in detail, shows that this mutation has at least two different and distinct origins, indicating that this mutation is selected independently of geographical location by *Ae. aegypti* populations [60].

DOI: http://dx.doi.org/10.5772/intechopen.1009757

4.1.1 What should we do to handle the issue of IR

New insecticides need to be introduced in the field. Insecticides based on natural products, essential oils (EO), can effectively replace current neurotoxic insecticides. These insecticides should have different modes of action to avoid cross-resistance. Cross-resistance is a common phenomenon in insects, deliberated by resistance developed against pyrethroids in DDT-resistant malaria vectors, as organochlorines (DDT) have a similar mode of action as pyrethroids, both targeting *vgsc* [61].

Until new insecticides are introduced in the field, our key focus should be to maximize the effectiveness of currently available insecticides. This emphasizes determining lethal doses for each local vector population rather than solely adhering to preestablished WHO-recommended doses. We can also use different insecticides in the same locality to delay the development of resistance. This can be done by distributing a locality into different patches and selecting an insecticide to be applied.

IR monitoring is essential for keeping a check on IR development. It should be made mandatory to perform bioassays and look for molecular mechanisms selected by vector populations.

In 2016, scientists from all over the world united in Rio de Janeiro to discuss IR on a global scale. A platform with the name WIN (World Insecticides resistance Network) held its second meeting in Singapore. Such a consortium is necessary as it discusses all aspects of IR. They also discussed the potential for novel insecticides and brought industry, academia, agriculture, and major stakeholders to the same table to discuss strategies related to VBDs and IR in particular [25, 62].

4.1.2 Other issues in current and conventional vector control

4.1.2.1 Presence of more than one vector

In several parts of Africa, the major malaria vector is *An. gambiae*, which was successfully eradicated in most parts of the continent. However, the arrival of An. stephensi from Asia disturbed this success [63].

The same can happen in the case of dengue, leishmaniasis, and other infections having more than one vector. Therefore, the presence of more than one vector species can greatly affect vector control.

4.1.2.2 Ignoring vector behavior while designing an intervention

Vector behavior is of utmost importance in designing any vector control strategy. Behaviors like resting, mating, feeding, and egg laying are extremely important while working on any control intervention. Any intervention that does not address these behaviors would not be successful [64].

4.1.2.3 Absence of effective vector surveillance tools

Vector surveillance is as important as any other control intervention. Many developing countries stop vector surveillance when there is no disease outbreak. Vector surveillance must be done throughout the year with or without the presence of disease outbreaks. Vector species must be studied for competence through robust molecular biology tools.

An effective vector surveillance system can help us prepare for disease outbreaks [65].

4.1.2.4 Sustainability and cost-effectiveness of innovative control tools

Novel control interventions like Wolbachia, SIT, copepods, etc., have sustainability issues. The majority of these techniques require an active molecular biology setup, a mosquito facility capable of producing millions of mosquitoes weekly, and continuous releases, among others [5]. None of these interventions is cost-effective, keeping in mind that VBDs mostly affect low- and middle-income countries where very little is invested in the health sector.

4.1.2.5 Community engagement and health education

Release of "modified" insects can provoke an unwanted response in local communities. This happened in a local Brazilian neighborhood, which was selected for the "*Wolbachia*" intervention. The shopkeeper noticed that after the release of "new" mosquitoes, his insecticide sales increased manyfold. Scientists involved in the intervention had to rethink and release a strain of *Wolbachia* males carrying a mutation to tolerate insecticides used by the local population [66].

The local community must be informed and taken on board before introducing any control intervention. A project initiated by Fiocruz researchers in Brazil, with the name "10 minutos contra Aedes," has shown excellent results. In this method, locals are shown the premature and mature stages of mosquitoes' life cycle with a focus on locating breeding sites in their surroundings. They are then asked to spare 10 minutes from their weekly schedule and look for these breeding sites at their homes and backyards. This method can be very effective if adopted locally in other regions of the world [67].

4.1.2.6 Lack of an integrated approach in vector control

The development of a yellow fever vaccine was initially believed to halt disease transmission. While it effectively controlled yellow fever, on the other hand, it led to a reduction in funding for vector control programs. This, in turn, facilitated the spread of other *Aedes*-borne diseases and contributed to the expanded geographical range of the *Ae. aegypti* mosquito.

Similarly, early success against *Ae. aegypti* eradication was halted by the lack of coordination among neighboring countries of Latin America [discussed earlier].

Therefore, each intervention must be well-communicated among stakeholders, and steps must be taken to ensure an effective and well-integrated approach.

4.2 Climate change and global warming

Global warming and climate change, as they are affecting other sectors of life, are also helping the spread of VBDs. Increased average global temperatures have resulted in more precipitation, which ultimately results in more breeding places for mosquitoes to nourish [68, 69]. Moreover, higher temperatures also favor mosquito propagation [70]. The recent spread of vector populations beyond their native range is also fueled by climate change. This is evident by the recent cases of chikungunya in Southern Europe [71].

Both active and passive surveillance should be guaranteed to stop the spread of VBDs and control them in regions where they are already present. If not, vector populations are expected to spread to all regions of the world, as predicted by Kraemer (**Figure 2**) [72].

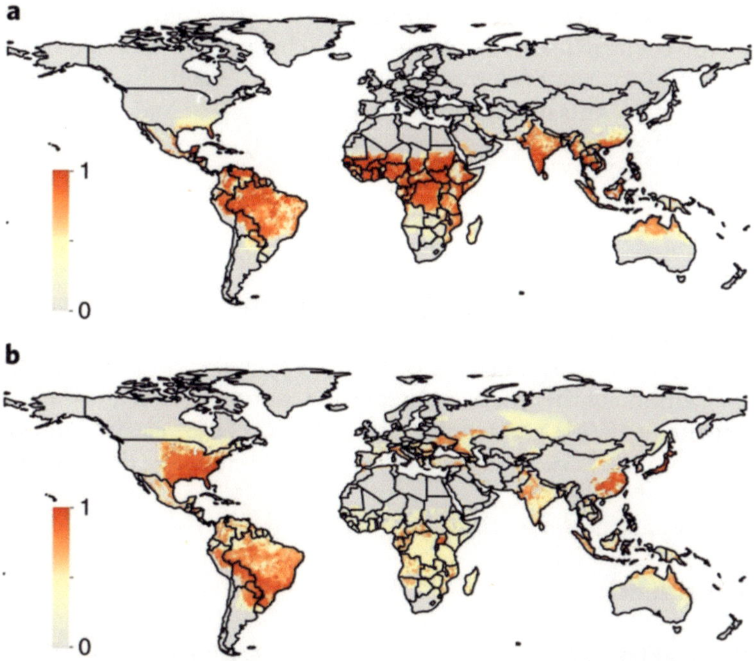

Figure 2.
Predicted future spread of (a) Aedes aegypti *and (b)* Aedes albopictus. *Adopted from Kraemer 2019.*

4.3 Unplanned urbanization and increased global connectivity

Among other issues that contribute to the global spread of VBDs is unplanned urbanization. An increase in global population is directly proportional to urbanization. Many rural areas are being transformed into urban settings with an increased demand for water storage. This favors an increase in stagnant water, which is ideal for the breeding of certain vectors like *Ae. aegypti* [73, 74].

No vector species can spread on its own, as their flight or self-distribution ability is very limited. Therefore, their spread is dependent on humans' moments and global trade. It is evident by the direct association of tire trade and the spread of dengue vectors. Historically, *Ae. aegypti* traveled from Africa to the New World through human migration and the stored water they brought with them for a sea voyage [75].

Community engagement, health education, and active airport screening should be guaranteed to tackle the global spread of VBDs through urbanization and global trade.

4.4 Lack of political will and conflict

As *Ae. aegypti* was eradicated in various Latin American countries, the lack of investment through governments in these countries and inactive border control led to the reintroduction of vector populations. Despite warnings from Soper, no investment was made in the area of vector control [16]. Similarly, malaria was almost eliminated from Pakistan, but the political situation in the border areas between Pakistan and Afghanistan and a huge influx of refugees from war-hit Afghanistan affected vector control in both countries.

Several African countries continue to struggle with the control of malaria and other vector-borne diseases due to persistent challenges. Political instability and unrest significantly impede progress by disrupting essential vector control and surveillance activities. This disruption creates a breeding ground for these diseases to spread and wreak havoc on affected communities [76].

5. Conclusions

Vector control faces several challenges, ranging from insecticide resistance to global warming. The potential of vector populations to adapt to extreme temperatures and their shorter life span with a huge progeny makes it extremely difficult to control and/or eliminate them. The only solution, as we see so far, is the Integrated Vector Management (IVM) and One Health. All sectors of society, along with political leadership, must be brought on the same page to make the dream of vector control a reality.

Acknowledgements

The authors acknowledge the Head of Department, Department of Biosciences, COMSATS University Islamabad, for his guidance and support. We also thank the Department of Parasite Biology, IOC (Fiocruz), for making important research articles available.

Conflict of interest

The authors declare no conflict of interest.

Author details

Rafiur Rahman[1,2]*, Naeem Ullah[3] and Ijaz Ali[1]

1 Department of Biosciences, COMSATS University Islamabad, Pakistan

2 Oswaldo Cruz Institute, Fiocruz, Rio de Janeiro, Brazil

3 Department of Community Medicine, Swat Medical College, Swat, Pakistan

*Address all correspondence to: rafitipu960@gmail.com

References

[1] Golding N et al. Integrating vector control across diseases. BMC Medicine. 2015;**13**:1-6

[2] UNDRR. Vector borne diseases (VBD) (Human). 2024. Available from: https://www.undrr.org/understanding-disaster-risk/terminology/hips/bi0023#:~:text=Vector%2Dborne%20diseases%20account%20for,more%20than%20700%2C000%20deaths%20annually

[3] Chilakam N et al. Economic burden of mosquito-borne diseases in low-and middle-income countries: Protocol for a systematic review. JMIR Research Protocols. 2023;**12**(1):e50985

[4] WHO. Updated WHO guidance for controlling vector-borne diseases through indoor residual spraying; 2024

[5] Wilson AL et al. The importance of vector control for the control and elimination of vector-borne diseases. PLoS Neglected Tropical Diseases. 2020;**14**(1):e0007831

[6] Ross R. On some peculiar pigmented cells found in two mosquitos fed on malarial blood. British Medical Journal. 1929;**1897**(2):1786

[7] Watson M. The prevention of malaria in the Federated Malay States. Liverpool: John Murray; 1921

[8] Le Prince JA, Orenstein AJ. Mosquito control in Panama; the eradication of malaria and yellow fever in Cuba and Panama. New York/London: Putnam; 1916

[9] Shousha AT. Species-eradication: The eradication of anopheles gambioe from upper Egypt, 1942-1945. Bulletin of the World Health Organization. 1948;**1**(2):309

[10] Camargo SD. History of *Aedes aegypti* eradication in the Americas. Bulletin of the World Health Organization. 1967;**36**(4):602

[11] Busvine JR. Disease transmission by insects: Its discovery and 90 years of effort to prevent it. New York/Berlin: Springer; 1993

[12] Kumar V et al. Field trial of an ecological approach for the control of Phlebotomus argentipes using mud & lime plaster. The Indian Journal of Medical Research. 1995;**101**:154-156

[13] Wheeler CM. Control of typhus in Italy 1943-1944 by use of DDT. American Journal of Public Health and the Nations Health. 1946;**36**(2):119-129

[14] Williams LL Jr. Malaria eradication in the United States. American Journal of Public Health and the Nations Health. 1963;**53**(1):17-21

[15] Soper FL. The elimination of urban yellow fever in the Americas through the eradication of Aedes aegypti. American Journal of Public Health and the Nations Health. 1963;**53**(1):7-16

[16] Soper ME. The Relationship between Personal Collections and the Selection of Cited References. University of Illinois at Urbana-Champaign; 1972

[17] Oswaldo Cruz Foundation Library. Biblioteca de Manguinhos

[18] Nájera JA, González-Silva M, Alonso PL. Some lessons for the future from the global malaria eradication programme (1955-1969). PLoS Medicine. 2011;**8**(1):e1000412

[19] Molineaux L, Gramiccia G. The Garki Project. Research on the Epidemiology and Control of Malaria in the Sudan Savanna of West Africa. Geneva: World Health Organization; 1980

[20] Oxborough RM. Trends in US President's malaria initiative-funded indoor residual spray coverage and insecticide choice in sub-Saharan Africa (2008-2015): Urgent need for affordable, long-lasting insecticides. Malaria Journal. 2016;**15**(1):146

[21] Bhatt S et al. The effect of malaria control on plasmodium falciparum in Africa between 2000 and 2015. Nature. 2015;**526**(7572):207-211

[22] Lowe R et al. The Zika virus epidemic in Brazil: From discovery to future implications. International Journal of Environmental Research and Public Health. 2018;**15**(1):96

[23] Vasconcelos PF, Monath TP. Yellow fever remains a potential threat to public health. Vector-Borne and Zoonotic Diseases. 2016;**16**(8):566-567

[24] Seventeenth meeting of the WHO Vector Control Advisory Group: Meeting report. 3-6 October 2022

[25] Corbel V et al. Tracking insecticide resistance in mosquito vectors of arboviruses: The worldwide insecticide resistance network (WIN). PLoS Neglected Tropical Diseases. 2016;**10**(12):e0005054

[26] Soloneski S, Larramendy M. Integrated Pest Management and Pest Control: Current and Future Tactics. Croatia: BoD–Books on Demand; 2012

[27] Hertlein MB et al. A review of spinosad as a natural product for larval mosquito control. Journal of the American Mosquito Control Association. 2010;**26**(1):67-87

[28] Scholte E-J et al. Entomopathogenic fungi for mosquito control: A review. Journal of Insect Science. 2004;**4**(1):19

[29] Guidi V et al. Dynamics of bacillus thuringiensis var. israelensis and Lysinibacillus sphaericus spores in urban catch basins after simultaneous application against mosquito larvae. PLoS One. 2013;**8**(2):e55658

[30] Soumare M, Cilek J. The effectiveness of Mesocyclops longisetus (Copepoda) for the control of container-inhabiting mosquitoes In residential Environments1. Journal of the American Mosquito Control Association. 2011;**27**(4):376-383

[31] Caputo B, Ienco A, Cianci D, Pombi M, Petrarca V, Baseggio A, et al. The "Auto-Dissemination" Approach: A Novel Concept to Fight *Aedes albopictus* in Urban Areas. PLOS Neglected Tropical Diseases. 2012;**6**(8):e1793. DOI: 10.1371/journal.pntd.0001793

[32] Devine GJ et al. Using adult mosquitoes to transfer insecticides to Aedes aegypti larval habitats. Proceedings of the National Academy of Sciences. 2009;**106**(28):11530-11534

[33] Mian L, Mulla M. Residue Reviews. New York, NY, USA: Springer; 1982

[34] Moreira LA et al. A Wolbachia symbiont in Aedes aegypti limits infection with dengue, chikungunya, and plasmodium. Cell. 2009;**139**(7):1268-1278

[35] Aly AS, Vaughan AM, Kappe SH. Malaria parasite development in the mosquito and infection of the mammalian host. Annual Review of Microbiology. 2009;**63**(1):195-221

[36] Frentiu FD et al. Limited dengue virus replication in field-collected Aedes aegypti mosquitoes infected with

Wolbachia. PLoS Neglected Tropical Diseases. 2014;**8**(2):e2688

[37] McGraw EA, O'neill SL. Beyond insecticides: New thinking on an ancient problem. Nature Reviews Microbiology. 2013;**11**(3):181-193

[38] Turley AP et al. Wolbachia infection reduces blood-feeding success in the dengue fever mosquito, Aedes aegypti. PLoS Neglected Tropical Diseases. 2009;**3**(9):e516

[39] Blagrove MS et al. Wolbachia strain w Mel induces cytoplasmic incompatibility and blocks dengue transmission in Aedes albopictus. Proceedings of the National Academy of Sciences. 2012;**109**(1):255-260

[40] Moretti R, Calvitti M. Male mating performance and cytoplasmic incompatibility in a wPip Wolbachia trans-infected line of Aedes albopictus (Stegomyia albopicta). Medical and Veterinary Entomology. 2013;**27**(4):377-386

[41] Hoffmann AA et al. Successful establishment of Wolbachia in Aedes populations to suppress dengue transmission. Nature. 2011;**476**(7361):454-457

[42] Patterson R et al. Suppression and elimination of an island population of Culex pipiens quinquefasciatus with sterile males. Science. 1970;**168**(3937):1368-1370

[43] Hannon GJ. RNA interference. Nature. 2002;**418**(6894):244-251

[44] Whyard S et al. Silencing the buzz: A new approach to population suppression of mosquitoes by feeding larvae double-stranded RNAs. Parasites & Vectors. 2015;**8**:1-11

[45] Harris AF et al. Field performance of engineered male mosquitoes. Nature Biotechnology. 2011;**29**(11):1034-1037

[46] Alphey L. Genetic control of mosquitoes. Annual Review of Entomology. 2014;**59**(1):205-224

[47] Sippy R et al. Ingested insecticide to control Aedes aegypti: Developing a novel dried attractive toxic sugar bait device for intra-domiciliary control. Parasites & Vectors. 2020;**13**:1-11

[48] Naranjo DP et al. Evaluation of boric acid sugar baits against Aedes albopictus (Diptera: Culicidae) in tropical environments. Parasitology Research. 2013;**112**:1583-1587

[49] Fulcher A et al. Attractive toxic sugar baits mixed with pyriproxyfen sprayed on plants against adult and larval Aedes albopictus (Diptera: Culicidae). Journal of Medical Entomology. 2014;**51**(4):896-899

[50] Revay EE et al. Control of Aedes albopictus with attractive toxic sugar baits (ATSB) and potential impact on non-target organisms in St. Augustine, Florida. Parasitology Research. 2014;**113**:73-79

[51] Raizada S, Mala S, Shankar A. Vector borne disease outbreak prediction by machine learning. In: 2020 International Conference on Smart Technologies in Computing, Electrical and Electronics (ICSTCEE). Bengaluru, India: IEEE; 2020. pp. 213-218. DOI: 10.1109/ICSTCEE49637.2020.9277286

[52] Blanford JI. Managing vector-borne diseases in a geoAI-enabled society. Malaria as an example. Acta Tropica. 2024;**260**:107406. DOI: 10.1016/j.actatropica.2024.107406. ISSN 0001-706X

[53] da Silva Motta D, Badaró R, Santos A, Kirchner F. Use of Artificial Intelligence on the Control of Vector-Borne Diseases [Internet]. Vectors and Vector-Borne Zoonotic Diseases. IntechOpen; 2019. DOI: 10.5772/intechopen.81671

[54] Organization W.H. Global Report on Insecticide Resistance in Malaria Vectors: 2010-2016; 2018

[55] Expert Committee on Insecticides: Seventh report of a meeting held in Geneva from 10 to 17 July 1956

[56] Brown A. The challenge of insecticide resistance. Bulletin of the ESA. 1961;7(1):6-19

[57] Hemingway J et al. The molecular basis of insecticide resistance in mosquitoes. Insect Biochemistry and Molecular Biology. 2004;**34**(7):653-665

[58] Ingham VA et al. A sensory appendage protein protects malaria vectors from pyrethroids. Nature. 2020;**577**(7790):376-380

[59] Scott JG. Life and death at the voltage-sensitive sodium channel: Evolution in response to insecticide use. Annual Review of Entomology. 2019;**64**(1):243-257

[60] Cosme LV et al. Evolution of kdr haplotypes in worldwide populations of Aedes aegypti: Independent origins of the F1534C kdr mutation. PLoS Neglected Tropical Diseases. 2020;**14**(4):e0008219

[61] Chadwick PR, Bowron MJ. An example of cross-resistance to pyrethroids in DDT-resistant Aedes aegypti. Pesticide Science. 1977;**8**(6):618-624

[62] Corbel V et al. Second WIN International Conference on "Integrated Approaches and Innovative Tools for Combating Insecticide Resistance in Vectors of Arboviruses", October 2018, Singapore. New York, USA. Berlin, Germany: Springer; 2019

[63] Sinka M et al. A new malaria vector in Africa: Predicting the expansion range of Anopheles stephensi and identifying the urban populations at risk. Proceedings of the National Academy of Sciences. 2020;**117**(40):24900-24908

[64] Lima-Camara TN et al. Effects of insemination and blood-feeding on locomotor activity of Aedes albopictus and Aedes aegypti (Diptera: Culicidae) females under laboratory conditions. Parasites & Vectors. 2014;**7**:1-8

[65] Leandro A, Maciel-de-Freitas R. Development of an integrated surveillance system to improve preparedness for arbovirus outbreaks in a dengue endemic setting: Descriptive study. JMIR Public Health and Surveillance. 2024;**10**:e62759

[66] Maciel-de-Freitas R. Listen to the shopkeeper. eLife. 2023;**12**:e87366

[67] Valle D. Campanha 10 Minutos Contra a Dengue-módulo 8. Brazil: Centro de Formação em Ciências da Saúde Dr. José Maria de Magalhães Neto; 2017

[68] Ma J et al. Climate change drives the transmission and spread of vector-borne diseases: An ecological perspective. Biology. 2022;**11**(11):1628

[69] Paz S. Effects of climate change on vector-borne diseases: An updated focus on West Nile virus in humans. Emerging Topics in Life Sciences. 2019;**3**(2):143-152

[70] Caminade C, McIntyre KM, Jones AE. Impact of recent and future climate change on vector-borne diseases. Annals of the New York Academy of Sciences. 2019;**1436**(1):157-173

[71] Rocklöv J et al. Using big data to monitor the introduction and spread of chikungunya, Europe, 2017. Emerging Infectious Diseases. 2019;**25**(6):1041

[72] Kraemer MU et al. Past and future spread of the arbovirus vectors Aedes aegypti and Aedes albopictus. Nature Microbiology. 2019;**4**(5):854-863

[73] Gubler DJ. Dengue, urbanization and globalization: The unholy trinity of the 21st century. Tropical Medicine and Health. 2011;**39**(4 Suppl):S3-S11

[74] Ibáñez-Justicia A. Pathways for introduction and dispersal of invasive Aedes mosquito species in Europe: A review. Journal of the European Mosquito Control Association. 2020;**38**:1-10

[75] Powell JR, Gloria-Soria A, Kotsakiozi P. Recent history of Aedes aegypti: Vector genomics and epidemiology records. Bioscience. 2018;**68**(11):854-860

[76] Abdulwahab A et al. Climate change, vector-borne diseases, and conflict: Intersecting challenges in vulnerable states. Journal of Infectious Diseases and Epidemiology. 2024;**10**:326

Chapter 2

Physiological Roles of Chemosensory Receptors in Lepidoptera

Dingze Mang

Abstract

Lepidoptera is one of the largest orders within the class Insecta, encompassing numerous species that are major pests in agriculture and forestry. These insects rely on a highly specialized chemosensory system to detect environmental chemical cues, which are essential for regulating key behaviors such as foraging, mating, oviposition, and predator avoidance. In the past two decades or so, substantial progress has been made in elucidating the functions of chemosensory receptors in Lepidoptera. This chapter summarizes advances in understanding the physiological roles of chemosensory receptors, with a particular focus on odorant receptors (ORs), gustatory receptors (GRs), ionotropic receptors (IRs), odorant-binding proteins (OBPs), and chemosensory proteins (CSPs). ORs expressed in the antennae detect sex pheromones and plant volatiles with high specificity, activating neural signaling pathways that mediate mating and host-seeking behaviors. GRs located in the maxillary palps and tarsi are involved in the perception of sugars, bitter compounds, and oviposition stimulants, thereby influencing feeding and oviposition preferences. Moreover, OBPs and CSPs serve as molecular carriers, facilitating the efficient transport of hydrophobic odorant molecules. A comprehensive understanding of the physiological roles of these chemosensory receptors provides a foundation for developing environmentally friendly pest management strategies.

Keywords: chemosensory neurons, gustation, olfaction, physiology, behavior

1. Introduction

Throughout long-term evolution, insects have developed highly specialized and extremely sensitive chemosensory systems, which play a fundamental role in mediating key behaviors involved in intra- and inter-specific interactions, as well as multitrophic relationships. These behaviors are crucial for the survival, reproduction, and evolutionary success of insect species. Insects utilize their finely tuned chemoreceptors to detect and discriminate among a wide array of chemical cues, enabling them to perform essential activities such as foraging, feeding, mating, and selecting suitable oviposition sites. The chemosensory signal transduction pathways in insects involve multiple proteins, among which olfactory receptors (ORs) and gustatory receptors

(GRs) are of particular importance. These receptors, typically composed of approximately 400 amino acid residues and characterized by seven transmembrane domains, bind to environmental chemical compounds and transmit the resulting signals to the central nervous system (CNS). The activation of specific neural circuits subsequently modulates insect behaviors in response to external chemical stimuli [1–3].

The Lepidoptera represents one of the most diverse groups within the Insecta. The chemosensory system of Lepidopteran plays a pivotal role in mediating essential life processes, including foraging, mating, oviposition, and defense. Elucidation of the mechanisms underlying chemical communication in Lepidoptera is fundamental for advancing our understanding of how these insects sustain populations, ensure survival, and achieve successful reproduction. Most Lepidopteran species, particularly moths, rely on species-specific sex pheromone communication and the precise recognition of host plant volatiles to locate appropriate hosts, facilitate mating, and establish reproductive isolation. Chemosensory receptors are central to these processes, serving as the fundamental functional units through which Lepidopteran insects perceive and respond to chemical cues from their environment [4, 5]. These receptors are broadly categorized into two types: olfactory receptors, located within olfactory sensilla, and gustatory receptors (also referred to as contact chemoreceptors), located within gustatory sensilla. Olfactory receptors detect volatile secondary metabolites emitted by host plants, thereby enabling host localization. Upon physical contact with the plant, gustatory receptors allow the insect to assess both the qualitative and quantitative aspects of plant-derived nutrients, feeding stimulants, and deterrents, thus facilitating informed behavioral decisions [6, 7].

In the past two decades or so, substantial progress has been made in the functional characterization of chemosensory receptors, driven by the rapid development of genomic and neurobiological technologies. This chapter focuses on the physiological functions of chemosensory receptors in Lepidopteran. First, a comprehensive overview of current knowledge regarding the structural organization and functional roles of the chemosensory system will be provided, followed by a summary of the perception mechanisms employed by olfactory and gustatory receptors. Second, recent advances in molecular and physiological studies of chemosensory receptors will be discussed. Finally, the methodology for chemosensory receptor function will be described.

2. Chemosensory system in Lepidoptera

2.1 Chemosensory organs

Chemoreception is crucial for insects to select suitable hosts, locate nutrient-rich foods, find mates, and avoid consuming toxic substances. Specialized chemoreceptors, such as gustatory receptors (GRs) and olfactory receptors (ORs), enable insects to detect chemicals over short and long distances, thereby triggering appropriate behavioral responses to these stimuli [8]. Such behaviors depend on the various chemosensory sensilla that are present on the antennae, mouthparts (including labrum, labium, maxillae, mandibles, and hypopharynx), legs, wings, and ovipositors (**Figure 1A-C**) [9, 10].

In insects, the sensillum is a simple structure of the epidermis and usually acts as a receptor for the perception of a variety of environmental stimuli [11]. Morphological and physiological characteristics of sensilla are involved in the

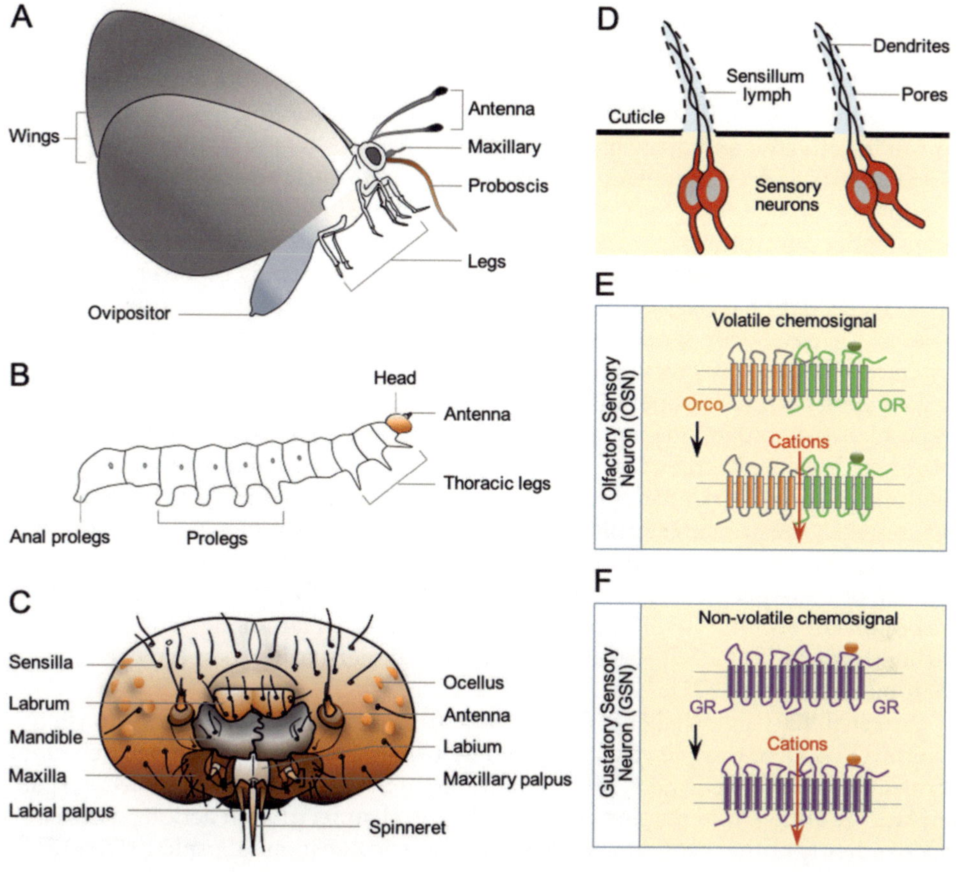

Figure 1.
Lepidoptera chemosensory organs on adults (A) and (B-C) larvae. Each sensory organ consists of several sensilla, which expresses chemosensory receptors (D). Model of mechanism of olfactory receptor complex (E), consisting of the co-receptor (Orco) and an ORx, and gustatory receptor (F).

evolutionary development of the insect chemosensory system [12]. Based on their wall structure and external appearance, sensilla are classified into different morphological types. The common types of sensilla in Lepidoptera are basiconic sensilla, trichoid sensilla, and coeloconic sensilla, each involved in different aspects of chemical detection [13]. Lepidopteran insects have uniporous gustatory sensilla (sensory organs) and multiporous olfactory sensilla. Gustatory sensilla can be broadly distributed into the mouthparts, antennae, wings, and ovipositor of the insects, where they are involved in eliciting various behaviors (**Figure 1A-C**) [14–17]. These sensilla styloconica have four gustatory receptor neurons (GRNs) and one mechanosensory neuron, which are involved in sensing taste and other related response [18]. A typical GRN has gustatory receptors (GRs) on its dendritic end with a specific affinity toward a diverse range of ligands such as sugars, salt (NaCl), inositol, and plant alkaloids. The olfactory sensilla generally contain specialized structures, including multiple cuticular pores, facilitating the detection of odor molecules [19]. The pores in the sensillum wall are connected to pore-tubules suspended in the sensillum lymph, which might serve as a route for the odor molecules to reach the odorant receptors (ORs) in the dendrites of the olfactory sensory neurons (OSNs) [20–22]).

2.2 Olfactory perception mechanism

The olfactory signal transduction mechanisms in Lepidopteran insects exhibit notable similarities to those characterized in *Drosophila*. Sodium fluoride (NaF), an G protein activator, and diacylglycerol analog 1-oleoyl-2-acetyl-sn-glycerol (DOG), which mimics endogenous diacylglycerol (DAG), both elicit neural impulses in the antennae of *Bombyx mori* that are comparable to those induced by sustained, low-intensity odorant stimulation. These compounds activate olfactory sensory neurons (OSNs) in both male and female moths, resulting in the opening of ion channels. Signal transduction processes are involved in pheromone-sensitive neurons in males and general odor-sensitive neurons in females [23]. In Lepidoptera, G protein α-subunits mediate olfactory signal transduction *via* interactions with G protein-coupled receptors (GPCRs), operating through two distinct second messenger pathways: cyclic adenosine monophosphate (cAMP) and inositol trisphosphate (IP3). In *B. mori*, two distinct olfactory receptors are specifically expressed in a pair of adjacent pheromone-sensitive neurons located in the male antennae. One neuron is tuned to bombykol, the primary sex pheromone, and the other to bombykal, its oxidized derivative. These two receptors are co-expressed, and their activation leads to the opening of non-specific cation channels (**Figure 1D** and **E**) [24]. The gene encoding the bombykol-specific GPCR is selectively expressed in pheromone-sensitive neurons during the pupal and adult stages of male moths, underscoring its specialized role in pheromone detection [25].

Sato et al. [26] and Wicher et al. [27] concurrently published studies proposing that the ORx–Orco complex functions as a non-selective, odorant-gated ion channel permeable to Ca^{2+}, Na^{+}, and K^{+} ions. According to Sato et al., odorant molecules can activate the ORx–Orco complex and open the ion channel to elicit action potentials independently of the second messenger cyclic adenosine monophosphate (cAMP). In contrast, Wicher et al. proposed a dual mechanism: at low odorant concentrations, and G protein signaling induces cAMP production, which subsequently activates Orco and promotes cation influx to generate action potentials; at higher concentrations, however, ion channel activation occurs directly, without the involvement of second messengers. Subsequent studies have demonstrated that Orco activation is also modulated by protein kinase C (PKC) and calmodulin (CaM) [26, 27].

A previous study demonstrated that pheromone signal transduction critically depends on the presence of sensory neuron membrane protein 1 (SNMP1) [28]. *In situ* hybridization and immunohistochemical analyses revealed that SNMP1 is typically co-expressed with the pheromone receptor (PR)-Orco complex in olfactory sensory neurons (OSNs) located in trichoid sensilla. When pheromones are delivered to the vicinity of the OSN membrane by pheromone-binding proteins (PBPs), SNMP1 is thought to induce conformational changes in PBPs, either by modulating the local pH or through direct interaction, thereby facilitating the release of the bound pheromones. SNMP1 next transfers the pheromone molecules to the adjacent PR-Orco complex, leading to its activation, cation influx, and initiation of downstream signal transduction [29].

In moths, the transduction of sex pheromone signals mediated by odorant receptors (ORs) also involves a G protein-coupled signaling cascade. Upon pheromone stimulation, antennal neurons exhibit elevated levels of inositol 1,4,5-trisphosphate (IP3) and generate electrophysiological responses. Activation of G proteins by pheromones initiates a signaling pathway involving β-type phospholipase C (PLC), which catalyzes the hydrolysis of phosphatidylinositol 4,5-bisphosphate (PIP2) into IP3 and diacylglycerol (DAG). The increase in IP3 activates cation channels on the membrane of OSNs, resulting in a Ca^{2+}-dependent inward current [30]. Concurrently, the accumulation of DAG activates protein kinase C (PKC), which in turn activates additional cation channels, including the Orco non-selective cation channel, thereby generating a sustained inward current [30]. This signaling cascade amplifies the sensitivity of male moths to sex pheromones, ensuring effective detection even at extremely low pheromone concentrations in the environment and enabling precise localization of conspecific females.

2.3 Gustatory perception mechanism

Insect feeding behavior depends on the integrated effects of gustatory and mechanical stimuli. For gustatory receptors (GRs) to respond to taste cues, direct contact with the substrate is required to encode mechanical information. In Lepidoptera, each of the two pairs of styloconic sensilla located on the maxillary galea contains four gustatory receptor neurons (GRNs). Some of these GRNs are tuned to detect feeding stimulants present in plant sap, while others respond to feeding deterrents. Gustatory signals are integrated and evaluated by the central nervous system, ultimately influencing host plant selection and feeding behavior [31, 32]. In the silkworm (*B. mori*), for instance, the four GRNs in the lateral styloconic sensillum are responsive to sucrose, myo-inositol, fructose, and salts, respectively. The GRNs in the medial styloconic sensillum detect deterrents, water, salts, and acids. Upon stimulation, interactions among neural impulses generated by these GRNs collectively determine feeding behavior [33, 34]. A recent study employing the silkworm-mulberry feeding model investigated the neural mechanisms underlying host plant recognition, proposing a "dual-factor host recognition mechanism" [35]. According to this model, the initial step of feeding is mediated by the maxillary palps (MP), which serves as highly sensitive chemosensory organs. During palpation of the leaf margin, the MP detects trace amounts of specific compounds, allowing pre-ingestive host plant recognition [36–38]. In *B. mori*, GRNs in the MP are highly sensitive to three key compounds found in mulberry leaves: isoquercitrin, β-sitosterol, and chlorogenic acid. Only when all three compounds are detected in combination does the silkworm initiate biting behavior. In the subsequent feeding phase, the maxillary galea—characterized by lower sensitivity—primarily detects sucrose in the sap exuding from damaged leaf tissue, facilitating sustained feeding [35].

The binding of ligands to their corresponding receptors initiates signal transduction. In the tobacco hornworm (*Manduca sexta*) larvae, for instance, the eight pairs of GRs cells located on the mouthparts exhibit distinct response profiles. Temporal coding within the peripheral gustatory system allows for coordinated recognition of bitter compounds, such as salicin, caffeine, and aristolochic acid, by three pairs of bitter-sensitive cells distributed across the medial styloconic sensilla, lateral styloconic sensilla, and epipharyngeal sensilla [39]. The CNS is capable of discriminating gustatory inputs arising from different signaling pathways within the same GR cell [40]. For example, 10 mmol/L myo-inositol has been shown to counteract the inhibitory effects of 10 mmol/L caffeine [40]. Within the CNS, excitatory and inhibitory inputs are integrated and transmitted as electrical signals to the brain, ultimately influencing feeding decisions in lepidopteran insects.

It is important to note that insect taste signal transduction—at least in the case of fructose sensing—operates *via* direct activation of ionotropic receptors, rather than through the traditional G protein-coupled receptor (GPCR) pathways. One of the well-understood mechanisms of GR functioning is an influx of extracellular Ca^{2+} ions through cation-gated channels (**Figure 1D** and **F**). This influx generates the depolarization of the dendritic ends of a GRN, and thus, the signal is generated. BmGr9 and HarmGr4, both specific to D-fructose, show this specific type of downstream signaling [41, 42]. This represents a paradigm shift in our understanding of insect gustation, aligning the mechanism more closely with ligand-gated ion channel signaling, similar to that observed in insect olfactory receptors.

3. Function of chemosensory receptors in Lepidoptera

3.1 Function of olfactory receptors

During odorant recognition in insects, the peripheral olfactory system binds and transports odorant molecules, triggering the generation, transmission, and integration of electrical signals. These signals are conveyed *via* axons to the central nervous system, ultimately eliciting behavioral responses to environmental stimuli. This intricate recognition process is mediated by the coordinated action of various olfaction-related proteins, including odorant binding proteins (OBPs), odorant receptors (ORs), ionotropic receptors (IRs), chemosensory proteins (CSPs), odorant-degrading enzymes (ODEs), and sensory neuron membrane proteins (SNMPs).

3.1.1 Odorant-binding protein

Once odorant molecules enter the sensillum lymph of the olfactory sensilla, odorant-binding proteins (OBPs) are the first to interact with and bind the molecules for recognition. Serving as carriers, OBPs transport the odorants to the surface of the olfactory receptors [43]. Typically, OBPs are small, soluble proteins with a molecular weight ranging from 12 to 20 kDa [44], synthesized by supporting cells and secreted into the sensillum lymph. In Lepidoptera, OBPs are functionally and structurally categorized into four types: pheromone-binding proteins (PBPs), general odorant-binding proteins (GOBPs), antennal binding proteins x (ABPx), and other OBPs [45].

Pheromone-binding proteins (PBPs) primarily function to bind and transport female sex pheromone components. These proteins are predominantly male-specific or highly expressed in the antennae [45–48]. Structural analysis of OfurPBP2 from the Asian corn borer *Ostrinia furnacalis* revealed that, under acidic conditions, a seventh α-helix is formed, which facilitates the displacement and release of small ligands from the binding pocket [49]. Similarly, the three-dimensional structure of *EposPBP3*, a PBP from the tortricid moth *Epiphyas postvittana*, demonstrated its ability to bind two major sex pheromone components [50]. Most moth species possess at least three PBP genes, which differ in both function and importance in pheromone detection. In the fall webworm *Hyphantria cunea*, the three identified PBPs—HcunPBP1, HcunPBP2, and HcunPBP3—exhibit significantly higher binding affinities for aldehyde pheromone components than those for epoxide components [51]. Some Lepidopteran species possess four to five PBP genes [52]. TintPBP4 was found to be highly expressed in male moths and binds strongly to sex pheromone components in *Tryporyza intacta* [53]. These findings collectively suggest that PBPs generally contribute to the specific recognition of various sex pheromone components and play essential roles in pheromone perception. Nevertheless, the functional differentiation among PBPs and their potential interactions remain to be fully elucidated.

Lepidopteran insects possess two general odorant binding protein (GOBP) genes that are highly conserved across species. These genes were first identified in olfactory sensilla responsible for detecting plant volatiles and are therefore thought to play a key role in host plant odor perception [54]. *In vitro* ligand-binding assays have demonstrated that MvitGOBP1 and MvitGOBP2 from *Maruca vitrata* are capable of binding to 17 plant-derived volatile compounds include butyl butyrate [55]. Similarly, SlitGOBP2 from *Spodoptera litura* shows high binding affinity to volatile compounds such as nerolidol [56]. In addition to their role in plant volatile detection, GOBPs, particularly GOBP2, have also been shown to bind sex pheromone components in various moth species, suggesting their potential involvement in pheromone perception. Collectively, these findings suggest that OBPs in Lepidoptera are primarily involved in odor detection, while also contributing to other physiological functions, such as development, insecticide response, and resistance formation.

3.1.2 Odorant receptor

Odorant receptors (ORs) are capable of specifically recognizing odor molecules and converting their chemical signals into electrophysiological signals through selective ion channels. These receptors are considered key components in the olfactory signal transduction pathway [57]. ORs are expressed in olfactory sensory neurons (OSNs), typically consist of 400–450 amino acids, and contain seven transmembrane domains. In insects, ORs are structurally categorized into two types: the odorant receptor co-receptor (Orco), which is highly conserved and broadly expressed across species, and the conventional odorant receptors (ORx), which are highly diverse and species-specific. Orco and ORx are usually co-expressed in the same OSN and can form a heteromeric complex (ORx-Orco) *via* a conserved C-terminal domain, functioning as a ligand-gated ion channel [26, 27, 58, 59]. In Lepidoptera, ORs can be functionally divided into pheromone receptors (PRs), which primarily detect sex pheromone components, and general odorant receptors, which are responsible for the detection of plant volatiles.

The odorant receptor co-receptor (*Orco*) is highly conserved across diverse insect species, and typically, only a single *Orco* gene is present in the genome of a given insect [60]. Functional studies in various Lepidopteran species have demonstrated that *Orco* knockout results in a complete loss of olfactory function. In *Helicoverpa armigera*, CRISPR-Cas9-mediated disruption of *Orco* abolished electrophysiological responses to sex pheromones and numerous host plant volatiles, suggesting that *Orco*-dependent olfactory signaling is essential for pheromone communication, oviposition site selection, and host-seeking behavior in this species [61]. Similar results have been observed in *Spodoptera frugiperda*, where *Orco* knockout males failed to respond to sex pheromones and were incapable of successful mating. Furthermore, mutant females exhibited a marked reduction in attraction to their host plant, maize, and larvae showed significantly impaired ability to locate food sources [62].

Pheromone receptors (PRs) play a crucial role in the specific recognition of sex pheromones. Typically, pheromone-sensitive OSNs express only a single, highly specific PR [63]. Studies in moths have demonstrated that PRs responsible for detecting Type I sex pheromones form a distinct clade within the order Lepidoptera [64]. Most lepidopteran species possess multiple PR genes, each exhibiting a narrow tuning range, usually responding to only one or two pheromone components. Functional analyses further support the involvement of PRs in pheromone detection. For instance, knockout of the *HvirOR14* gene in *Heliothis virescens* did not affect mating rate or electrophysiological responses but resulted in prolonged mating duration [65].

In *Spodoptera littoralis*, disruption of the *SlitOR5* gene abolished male electrophysiological responses to the major pheromone component [66].

3.1.3 Ionotropic receptor

A class of receptor genes homologous to ionotropic glutamate receptors (iGluRs), termed ionotropic receptors (IRs), has been identified in the odor-sensitive antennal olfactory sensory neurons (OSNs) of *Drosophila melanogaster*. Notably, these IRs are expressed in OSNs distinct from those expressing odorant receptors (ORs) [67]. IRs have been shown to mediate the detection of various odorants, including phenylacetaldehyde, amines, and acids, and are also involved in thermosensation and hygrosensation. IRs can be broadly categorized into two subfamilies: divergent IRs and antennal IRs. Divergent IRs exhibit low sequence conservation across species, suggesting species-specific functions, whereas antennal IRs are more evolutionarily conserved and are likely present in all insect taxa. Among these, only the antennal IRs are relatively conserved within Diptera and Hymenoptera. To date, IRs have been identified in several insect orders, including Lepidoptera, Diptera, and Coleoptera.

In recent years, functional studies of IRs in Lepidoptera have begun to emerge. For instance, knockout of *MsIR8a* in the tobacco hornworm (*M. sexta*) resulted in a diminished avoidance response by females to fecal odor volatiles, indicating that *MsIR8a* plays an essential role in feces-avoidance oviposition behavior, which is critical for minimizing interspecific competition [68]. Similarly, deletion of *HarmIR8a* in the cotton bollworm led to a significant reduction or complete loss of acid sensitivity, suggesting that *HarmIR8a* is involved in mediating the detection of short-chain carboxylic acids and acetic acid that are attractive to this species [69]. In the turnip moth (*Agrotis segetum*), *AsIR75p.1* and *AsIR75q.1* have been shown to respond specifically to hexanoic acid and octanoic acid, respectively [70]. Our recent study further revealed that *HcIR68a* responds to environmental temperature fluctuations and transmits thermal signals, leading to the expression of capability peptides in *H. cunea* larvae [71]. Collectively, these findings indicate that research on IR functions in Lepidoptera is still in its early stages. To date, only a limited number of antennal IRs have been functionally characterized, and the roles of many other highly expressed antennal IR genes are yet to be elucidated.

3.1.4 Chemosensory protein

Chemosensory proteins (CSPs) represent a class of soluble proteins initially identified in the antennae of *D. melanogaster*, characterized by structural and functional properties distinct from those of OBPs [72]. These proteins are implicated in the detection of chemical stimuli and are abundantly expressed in the sensillar lymph of chemosensory organs, functioning analogously to OBPs. CSPs are expressed not only in chemosensory tissues but also in a variety of non-chemosensory organs, suggesting multifunctional roles beyond olfaction [73]. Certain CSPs are specifically expressed in the antennae and have been shown to bind plant volatiles and pheromonal compounds [74]. Others are highly expressed in non-olfactory tissues such as legs, thorax, abdomen, wings, and female reproductive glands [75]. Antennal-enriched CSPs are widely considered to be involved in odorant recognition. For example, SiCSP19 from *Sesamia inferens* binds strongly to six plant volatiles and three sex pheromone components or analogs, indicating its dual role in the detection of female sex pheromones and host plant cues [76]. Similarly, HarmCSP6 from *H. armigera* is believed to participate in female pheromone detection [77]. In addition to chemosensory functions, CSPs have been implicated in the immune response and resistance mechanisms to insecticides in Lepidoptera. In *B. mori*, the expression levels of several CSP genes were significantly upregulated following exposure to abamectin [78]. Likewise, treatment with λ-cyhalothrin resulted in time-dependent changes in CsCSP expression in both male and female *Conopomorpha sinensis* adults [79], suggesting a potential role for CSPs in insecticide metabolism and resistance. In summary, CSPs in Lepidopteran species exhibit multi-functionality, playing critical roles in olfactory perception and participating in a range of physiological processes.

3.1.5 Odorant-degrading enzyme

The first odorant-degrading enzyme (ODE) was identified in 1981 in the giant silk moth *Antheraea polyphemus*, where it was found to be highly expressed in the male antennae [80]. Subsequent studies revealed that ApolPDE is a carboxylesterase gene and accordingly classified it as a pheromone-degrading enzyme (PDE). In recent years, carboxylesterases (CXEs) in moths have been shown to exhibit distinct substrate degradation profiles. Individual esterases are capable of degrading both long-chain ester-type pheromone components and short-chain plant volatiles, typically displaying higher affinity for long-chain esters and faster catalytic efficiency for short-chain esters [81–83]. Furthermore, multiple ODE genes have been identified within a single species. In the diamondback moth (*Plutella xylostella*), for instance, two carboxyl/cholinesterases (PxylCCE16a and PxylCCE16c) are highly expressed in female antennae and are capable of degrading both acetate-type pheromone components and various ester-type plant volatiles, with a stronger affinity toward pheromone substrates [84, 85]. In the oriental fruit moth (*Grapholita molesta*), knockdown of four antennal-expressed CXE genes (GmolCXE1, GmolCXE5, GmolCXE14, and GmolCXE21) resulted in significantly reduced behavioral responses to ester-type pheromones and host plant volatiles [86].

Aldehyde oxidases (AOXs) are a class of enzymes that catalyze the oxidation of aldehydes to their corresponding carboxylic acids. Together with xanthine oxidoreductases (XORs), they comprise the molybdo-flavoenzyme (MFE) family [87]. In the diamondback moth (*Plutella xylostella*), an aldehyde oxidase (PxylAOX3) highly expressed in the antennae has been shown to oxidize aldehyde sex pheromones and their analogs, plant-derived aldehyde volatiles, and aldehyde-containing pesticides [85]. Similarly, in the greater wax moth (*Galleria mellonella*), GmelAOX plays a critical role in the metabolism of aldehyde odorant signal [88].

Cytochrome P450 enzymes are among the most critical detoxification enzymes in living organisms. The high expression levels of certain P450 members in the antennae suggest that they play an important role in the inactivation and clearance of odorant signals [89, 90]. However, P450 enzymes are intracellular and anchored to the membrane of the endoplasmic reticulum, indicating that they can degrade odorant molecules only after these compounds have entered the cell [91]. Consequently, their influence on olfactory sensitivity is likely indirect, potentially serving to protect membrane-bound receptors or to maintain normal cellular and physiological functions by metabolizing toxic substances that penetrate the cell.

3.1.6 Sensory membrane protein

Sensory neuron membrane protein (SNMP) was initially identified as an antenna-specific protein expressed in pheromone-sensitive olfactory sensory neurons (OSNs) of the Lepidopteran moth *Antheraea polyphemus* [92]. Lepidopteran species generally possess three SNMP genes. In *A. polyphemus*, SNMP1 is predicted to be a double transmembrane protein consisting of 519–525 amino acids with the post-translational modifications. It is initially synthesized as a 59 kDa polypeptide and processed into a 69 kDa form, with transmembrane domains at both termini [93, 94].

In *H. armigera*, HarmSNMP1 is specifically expressed in adult antennae, whereas HarmSNMP2 shows broad expression across adult antennae, tarsi, larval antennae, and mouthparts. HarmSNMP3 is specifically expressed in the larval midgut. The expression levels of HarmSNMP2 and HarmSNMP3 vary significantly depending on the host plant species, indicating their roles in feeding behavior [95]. Notably, deletion of HarmSNMP1 abolishes male responsiveness to sex pheromones and significantly reduces mating success with wild-type females. HarmSNMP1 is specifically involved in detecting long-chain (C16) pheromones, but not short-chain (C14) pheromones or plant volatiles [96]. Similarly, knockdown of BmSNMP1 in *B. mori* led to impaired mating behavior, indicating a key role for SNMP1 in sex pheromone detection [97].

3.2 Function of gustatory receptors in lepidopteran

Lepidopteran insects exhibit diverse life history strategies, and gustatory receptors (GRs) play a pivotal role in mediating their feeding behaviors. GRs in Lepidoptera are closely associated with host plant specificity and can even determine the feeding preferences of these insects. Therefore, comprehensive analyses of Lepidopteran GRs are essential for understanding their functional roles in insect evolution and ecological adaptation. Based on phylogenetic relationships and ligand specificity, GRs are categorized into four major subfamilies: sugar receptors, Gr43a-like receptors, bitter receptors, and CO_2 receptors.

3.2.1 Sugar gustatory receptors

The function of sugar receptors in insects has been extensively studied in *D. melanogaster*. Dahanukar et al. generated a *Gr5a*-deficient mutant line and demonstrated that this strain exhibited a complete loss of both electrophysiological and behavioral responses to trehalose, while responses to sucrose remained unaffected [98]. These results indicate that *Gr5a* is specifically involved in trehalose perception but not in sucrose detection. Subsequent studies revealed that *DmGr64a* is implicated in the perception of sucrose and maltotriose [99]. Furthermore, Jiao et al. showed that *DmGr64f* can functionally interact with *DmGr5a* to mediate trehalose detection and with *DmGr64a* to mediate sucrose detection. These findings suggest that the recognition of sugars such as sucrose in *D. melanogaster* requires the coordinated activity of multiple gustatory receptors (GRs) [100]. Fujii et al. further demonstrated that different sugar-sensitive sensilla in *D. melanogaster* express distinct combinations of sugar receptors, including *DmGr5a*, *DmGr61a*, *DmGr64a*, *DmGr64b*, *DmGr64c*, *DmGr64d*, *DmGr64e*, and *DmGr64f*. These receptors play essential roles and are expressed in diverse combinations across various types of sugar-sensing sensilla, enabling the detection of a wide range of sugar molecules [101].

To date, research on GRs in Lepidoptera remains limited. While the expression profiles of some GRs have been characterized, their physiological functions are still largely unexplored. BmGr4 has been identified as a receptor for sucrose and glucose, and it may serve as a chemosensor for nutritional compounds in both the midgut and brain [102]. BmGr6 and BmGr9 were found to be co-localized with neuronal cells in the brain, suggesting a potential role for BmGr6 in the regulation of feeding behavior in *B. mori* larva [103]. Additionally, BmGr6 has been implicated in the secretion of Bommo-myosuppressin (BMS), a neuropeptide that may influence feeding behavior [104]. In *H. armigera*, HarmGr10 is primarily utilized by larvae to detect low concentrations of sucrose, whereas HarmGr6 is predominantly used by adults to sense higher concentrations of sucrose and other saccharides. These differences likely underlie the observed variation in sugar selectivity and sensitivity between larval and adult sensory neurons [105].

3.2.2 Gr43a-like gustatory receptors

Miyamoto et al. reported that *D. melanogaster* Gr43a (DmGr43a) is specifically tuned to fructose [106]. This receptor is also expressed in the brain, where it modulates feeding behavior in starved flies. Fructose receptors in Lepidoptera were initially identified based on sequence similarity to DmGr43a [107]. Functional analyses demonstrated that when *B. mori* Gr9 (BmGr9) was heterologously expressed in *Xenopus* oocytes and HEK293T cells, fructose elicited extracellular calcium influx and non-selective cation currents, confirming that fructose serves as a ligand for BmGr [41]. Since then, fructose-responsive gustatory receptors (GRs) have been identified in several other insect species, including HarmGr4 in *H. armigera* [42, 108], PxGr43a-1 in *P. xylostella* [96], SlitGr8 in *Spodoptera litura* [109], and HcGr76 in *Hyphantria cunea* [110]. These receptors are orthologous to DmGr43a and are directly tuned to fructose without the need to form heteromeric complexes with other GRs. Notably, HcGr76 is also expressed in the larval midgut, where it plays a role in detecting dietary components and regulating the expression of peptide hormones involved in metabolic processes [110].

Phylogenetic analysis further supports that fructose-responsive GRs across insect species cluster within a single evolutionary branch, suggesting a high degree of conservation in both sequence and function. In *B. mori*, BmGr8 and BmGr10 have been shown to respond to inositol [111, 112]. Recent findings also indicated that some GRs within the sugar receptor clade can detect specific plant secondary metabolites. For instance, BmGr6 and BmGr9 respond not only to high concentrations of sucrose, glucose, and fructose, but also exhibit pronounced sensitivity to chlorogenic acid and isoquercitrin—two feeding stimulants in *B. mori* [113].

3.2.3 Bitter gustatory receptors

Functional studies on insect bitter taste receptors have predominantly focused on *D. melanogaster*. In this species, five gustatory receptors—DmGr47a, DmGr32a, DmGr33a, DmGr66a, and DmGr22e—are involved in the detection of strychnine [114–117]. Nicotine perception is mediated by DmGr10a, DmGr32a, and DmGr33a, while canavanine is detected by DmGr8a, DmGr66a, and DmGr98b [118]. Furthermore, DmGr33a, DmGr66a, and DmGr93a are responsible for sensing caffeine and umbelliferone [119–121], and DmGr28b contributes to the detection of saponin [122].

Numerous GRs have been identified in Lepidopteran insects, with the majority functioning as bitter receptors. Functional studies of these receptors are rapidly increasing. Unlike in *Drosophila*, a single Lepidopteran bitter GR can form a functional receptor in heterologous expression systems [41]. To date, no clear classification of broadly versus narrowly tuned bitter GRs has been established in Lepidoptera, although several receptors are capable of recognizing multiple bitter compounds. For example, RNAi-mediated knockdown of *PxutGr1* in the citrus swallowtail (*Papilio xuthus*) significantly reduced electrophysiological responses of specific tarsal sensilla to synephrine, as well as the oviposition preference for this compound [123]. In the silkworm, CRISPR-Cas9-mediated knockout of *BmGr66* abolished the insect's specific feeding behavior toward mulberry leaves [124]. Other silkworm GRs have also been functionally characterized: *BmGr16* and *BmGr53* respond to caffeine and coumarin, while *BmGr18* is involved in the detection of caffeine, pilocarpine, and coumarin [125]. *BmGr63* responds to the feeding stimulant isoquercetin [126], and *BmGr67* is activated by berberine [127]. In the diamondback moth (*P. xylostella*), *PxylGr34* has been shown to detect the plant hormone brassinolide and its analog 24-epi-brassinolide [128]. In the cabbage white butterfly (*Pieris rapae*), *PrapGr28* is tuned to sinigrin, a characteristic glucosinolate of cruciferous plants [129]. In *H. armigera*, Harm*Gr180* mediates the detection of coumarin, sinigrin, and strychnine [130]. Additionally, in the fall webworm (*H. cunea*), HcGr148 has been found to respond to quinine, a plant-derived feeding deterrent [131].

The diversification of GR families is closely linked to insect ecological niches and adaptive capacities. In many Lepidopteran species, bitter GRs have undergone extensive gene family expansion. For example, 55 bitter GRs resulting from a single lineage-specific expansion have been identified in the silkworm genome [107, 132]. In other lepidopteran pests such as *H. armigera*, *Spodoptera frugiperda*, and *Spodoptera exigua*, over 200 GRs have been reported. These receptors are potentially involved in regulating feeding behaviors and oviposition preferences [42, 109]. In *Heliconius melpomene*, 73 GRs have been identified, the majority of which arose from duplications of a common ancestral gene and are undergoing rapid gene family evolution [133]. Tandem duplications of GRs are widely observed across Lepidoptera, and the expansion in gene number may correlate with the diversification of angiosperms. This gene expansion likely facilitates the detection of host plant attractants and toxic secondary metabolites, thereby enhancing insect adaptability to a variety of host plants [134, 135].

3.2.4 CO_2 receptors

Carbon dioxide (CO_2) plays a significant role in the foraging behavior of both larval and adult Lepidoptera. For instance, larvae of the *H. armigera* exhibit a preference for food sources with elevated CO_2 levels [136]. Similarly, larvae of the lesser cornstalk borer (*Elasmopalpus lignosellus*) are attracted to CO_2 when locating food resources [137]. Many lepidopteran species feed on floral nectar, and it has been suggested that flowers emit relatively high concentrations of CO_2, which visiting insects may use as a foraging cue to locate nectar-rich blooms [138, 139]. In addition to foraging, CO_2 also influences reproductive behaviors. In nocturnal moths such as the cactus moth (*Cactoblastis cactorum*), oviposition behavior is modulated by CO_2 cues [140, 141]. Elevated CO_2 concentrations in the environment of its host plant, *Opuntia stricta*, significantly reduce the number of eggs laid on the cactus, highlighting the importance of CO_2 as an oviposition signal [142]. Similar findings have been reported in the tobacco hawkmoth, where increased CO_2 levels around host plants result in reduced oviposition by females [143]. Moreover, high CO_2 concentrations can interfere with the perception of sex pheromones in *H. armigera*, potentially impacting mating behaviors [144].

Significant progress has been made in understanding the mechanisms underlying CO_2 detection in insects since the first identification of CO_2 receptors in *Drosophila* in 2007 [145]. These advances have primarily focused on identifying of CO_2 receptor genes across diverse insect species, functionally characterizing these receptors, and mapping of the neural projection pathways of CO_2-sensitive neurons in the brain. In Lepidopteran insects, including *H. armigera* [146–148], *Mythimna separata* [149, 150], *H. assulta* [150, 151], *M. sexta* [152], and *B. mori* [107], three CO_2 receptor genes—*GR1*, *GR2*, and *GR3*—have been identified. In these species, CO_2-sensitive receptor neurons are predominantly localized within a specialized sensory structure known as the labial-palp pit organ (LPO).

Compared to *Drosophila* and mosquitoes, the functional characterization of CO_2 receptors in other insect species remains limited. In *H. armigera*, three putative CO_2 receptor genes—*HarmGR1*, *HarmGR2*, and *HarmGR3*—were individually expressed in Sf9 cells. Calcium imaging analysis revealed that only cells expressing *HarmGR3* exhibited a significant response to sodium bicarbonate solution [148]. In a subsequent study, the three receptors were expressed either individually or in various combinations (*HarmGR1* + *HarmGR2*, *HarmGR1* + *HarmGR3*, *HarmGR2* + *HarmGR3*, and *HarmGR1* + *HarmGR2* + *HarmGR3*) in *Xenopus* oocytes. Two-electrode voltage clamp recordings demonstrated that a detectable response to sodium bicarbonate solution occurred only when *HarmGR1* and *HarmGR3* were co-expressed [153]. These findings suggest that the *in vitro* functionality of CO_2 receptor genes in *H. armigera* remains controversial. When RNA interference (RNAi) was used to individually knock down three CO_2 receptor genes—*DvvGR1*, *DvvGR2*, and *DvvGR3*—in the western corn rootworm (*Diabrotica virgifera*), behavioral assays revealed that *DvvGR2* plays a critical role in larval orientation toward CO_2 [154].

4. Methodology for chemosensory receptor function

Methods for investigating the functions of chemosensory receptors are generally conducted using both *in vitro* and *in vivo* approaches. *In vitro* studies typically utilize heterologous expression systems using various cell lines, including HEK293 cells, insect Sf9 cells, and *Xenopus laevis* oocytes. Among these, the *Xenopus* oocyte expression system is the most widely used and well-established method in entomological research [41, 155]. *In vivo* studies commonly employ electrophysiological techniques or genetic tools such as the GAL4-UAS transgenic *Drosophila* empty neuron system, CRISPR-Cas9, and RNA interference (RNAi). Notably, the transgenic *Drosophila* empty neuron system has become a prevalent approach for functional characterization of chemosensory receptors [156–158].

4.1 Ca^{2+} imaging

Ca^{2+} imaging has been proven to be a valuable tool in the identification of receptors and their corresponding ligands [41, 112]. This approach involves heterologous expression of the receptor of interest in a suitable cell line, followed by stimulation with exogenous compounds. Activation of the receptor by a ligand typically triggers calcium influx from the extracellular space into the cytoplasm, raising intracellular calcium concentration. This change can be detected using calcium indicators, which have both a high affinity for calcium ions and fluorescent properties. Since the fluorescence intensity of these indicators is directly correlated with intracellular calcium levels, this method enables the effective characterization of receptor-ligand interactions.

4.2 Two-electrode voltage-clamp in *Xenopus laevis* oocytes

The *Xenopus laevis* oocyte expression system enables the functional characterization of target olfactory receptors by heterologous-expressing receptor genes in oocytes and measuring compound-induced membrane current changes using two-electrode voltage clamp (TEVC) recording. This approach has been widely applied to investigate the ligand-binding properties of olfactory receptors in insects from several orders, including Diptera, Coleoptera, Lepidoptera, and Hymenoptera [155, 159, 160]. Owing to its operational simplicity and high efficiency, this system is well-suited for high-throughput screening of receptor-ligand interactions. However, as ligand compounds are applied in solution rather than *via* natural odor delivery routes, the system does not fully replicate the physiological conditions of the insect olfactory environment, which may result in false positives in certain circumstances [59].

4.3 Electrophysiological recording

Electrophysiological assays are widely employed in studies of insect chemosensation. Chemosensory processes, including both olfactory and gustatory perception, involve the conversion of chemical stimuli into electrical signals. This signal transduction is typically accompanied by transmembrane ion fluxes, both intracellular and extracellular. Electrophysiology utilizes specialized electrical devices to detect and amplify these ionic signals, rendering them in a visualized format. Electrophysiological recordings are generally categorized into *in vivo* and *in vitro* approaches. In the context of gustatory research, *in vivo* techniques, such as tip recording and single sensillum recording (SSR), are commonly used to measure neural responses of insects to various metabolites. In a typical setup, the insect's head is excised, and one end of a reference silver electrode is inserted into the head capsule, with the other end connected to an amplifier, which is in turn linked to a data acquisition system. The glass recording electrode, filled with the test solution, is brought into contact with gustatory sensilla. Upon stimulation, gustatory receptor neurons (GRNs) generate electrical signals, which are processed and displayed as visible spike patterns. These recordings allow for the evaluation of gustatory receptor sensitivity to specific chemical compounds [113, 161, 162].

4.4 *Drosophila* empty neuron

The *Drosophila* empty neuron system employs the GAL4-UAS binary expression system to ectopically express target chemosensory receptor genes in specific olfactory sensory neurons. This allows for functional characterization of the receptors *via* single sensillum recording (SSR), enabling precise analysis of neuronal response profiles [59, 156]. Because *Drosophila* olfactory sensilla provide the necessary cellular environment and cofactors for olfactory signal transduction, this heterologous expression system offers a physiologically relevant context. Moreover, it can be readily integrated with behavioral assays, thereby providing a robust and intuitive platform for elucidating the *in vivo* function of candidate olfactory receptors.

5. Conclusion

Insects employ highly specialized chemosensory receptors to detect host-derived chemical cues, including extremely low concentrations of signals such as sex pheromones, thereby facilitating critical behaviors such as foraging, mating, oviposition, habitat selection, and host localization. Recent advances in the study of the molecular mechanisms underlying insect chemosensation have significantly enhanced our understanding of the chemosensory neural perception systems in insects. Lepidoptera represents one of the most diverse orders within the insect kingdom, comprising over 160,000 described species. Notably, Lepidopteran insects account for more than 70% of storage, agricultural, and forestry pests globally. The widespread and increasing resistance of these pests to conventional insecticides underscores the urgent need for the development of novel, environmentally sustainable pest management strategies. With the continuous development of innovative genetic manipulation technologies and analytical methods, knowledge of chemosensory systems, particularly in Lepidopteran species, is becoming increasingly comprehensive.

The application of chemosensory principles is anticipated to play an increasingly vital role in the development of novel strategies for pest management. Despite these advances, the functional characterization of numerous chemoreceptors in non-model insect species, particularly major agricultural pests beyond *Drosophila*, remains largely incomplete. In many cases, the functions of chemosensory receptors are still unknown. Pheromone-based strategies leveraging insect olfactory mechanisms have demonstrated considerable efficacy in pest monitoring and control. However, the potential utility of feeding and oviposition stimulants or deterrents—based on gustatory mechanisms—has yet to be fully explored and assessed. Looking ahead, it is essential to further strengthen research on the chemical communication between Lepidopteran insects and their environment, as well as their interactions with plant-derived information. A deeper investigation into the fundamental nature of insect behavioral responses will facilitate practical applications and open new avenues for the development of insect behavior-modulating agents.

By generating large-scale genomic and transcriptomic datasets, we can systematically compare the gene families, expression patterns, and functions of chemosensory receptors across different insect species, thereby revealing the functional diversity and evolutionary trajectories of these systems. Overall, advancing our understanding of chemosensory system and associated receptors not only sheds light on the mechanisms underlying chemical communication and host recognition and selection in insects but also provides critical insights into the broader mechanisms driving insect behavior and the evolutionary processes of chemosensation.

Acknowledgements

I would like to thank members of the Mang Lab for providing data. I also thank Dr. Hong Huat Hoh (Sunway Clinical Research Centre, Sunway Medical Centre, Malaysia) for English language editing and Dr. Endang R. Purba (Okinawa Institute of Science and Technology Graduate University, Japan) for assisting in drawing the figure. This work was supported in part by the Top-notch Talent Program of Hebei province (HY2024050018), Early-Career Scientists (21 K14859) from the Japan Society for the Promotion of Science, and the Excellent Youth Research Innovation Team of Hebei University (QNTD202501).

Conflict of interest

The author declares no conflict of interest.

Author details

Dingze Mang
College of Life Science, Institute of Life Science and Green Development,
Hebei University, Baoding, China

*Address all correspondence to: mang@hbu.edu.cn

References

[1] Benton R. Multigene family evolution: Perspectives from insect chemoreceptors. Trends in Ecology & Evolution. 2015;**30**(10):590-600

[2] Joseph RM, Carlson JR. *Drosophila* chemoreceptors: A molecular Interface between the chemical world and the brain. Trends in Genetics. 2015;**31**(12):683-695

[3] Robertson HM, Robertson ECN, Walden KKO, Enders LS, Miller NJ. The chemoreceptors and odorant binding proteins of the soybean and pea aphids. Insect Biochemistry and Molecular Biology. 2019;**105**:69-78

[4] Agnihotri AR, Roy AA, Joshi RS. Gustatory receptors in Lepidoptera: Chemosensation and beyond. Insect Molecular Biology. 2016;**25**(5):519-529

[5] Xu W. How do moth and butterfly taste?—Molecular basis of gustatory receptors in Lepidoptera. Insect Science. 2020;**27**(6):1148-1157

[6] Robertson HM, Warr CG, Carlson JR. Molecular evolution of the insect chemoreceptor gene superfamily in *Drosophila melanogaster*. Proceedings of the National Academy of Sciences. 2003;**100**(Suppl_2):14537-14542

[7] Hansson B, Wicher D. Chapter 2 - chemical ecology in insects. In: Zufall F, Munger SD, editors. Chemosensory Transduction. Academic Press; 2016. pp. 29-45

[8] Depetris-Chauvin A, Galagovsky D, Grosjean Y. Chemicals and chemoreceptors: Ecologically relevant signals driving behavior in *Drosophila*. Frontiers in Ecology and Evolution. 2015;**3**

[9] Zacharuk RY, Shields VD. Sensilla of immature insects. Annual Review of Entomology. 1991;**36**(1):331-354

[10] Dey S, Singh S, Chakraborty R. Surface ultrastructure of larval mouthpart sensilla of the muga silkmoth, *Antheraea assamensis*, an endemic species of north-East India. Microscopy Research and Technique. 2011;**74**(3):292-300

[11] Hartenstein V. 1.11 - Development of insect Sensilla. In: Gilbert LI, editor. Comprehensive Molecular Insect Science. Amsterdam: Elsevier; 2005. pp. 379-419

[12] Missbach C, Dweck HKM, Vogel H, Vilcinskas A, Stensmyr MC, Hansson BS, et al. Evolution of insect olfactory receptors. eLife. 2014;**3**:e02115

[13] Wang W, He P, Liu T, Jing X, Zhang S. Morphology and distribution of antennal Sensilla on *Spodoptera frugiperda* (Lepidoptera: Noctuidae) larvae and adults. Diversity [Internet]. 2023;**15**(9)

[14] Chapman RF. Chemoreception: The significance of receptor numbers. In: Berridge MJ, Treherne JE, Wigglesworth VB, editors. Advances in Insect Physiology. 16: Academic Press; 1982. pp. 247-356

[15] Störkuhl KF, Hofbauer A, Keller V, Gendre N, Stocker RF. Analysis of immunocytochemical staining patterns in the antennal system of *Drosophila melanogaster*. Cell and Tissue Research. 1994;**275**(1):27-38

[16] Dethier VG. The Hungry Fly: A Physiological Study of the Behavior Associated with Feeding. Oxford, England: Harvard U Press; 1976. 489 p

[17] Dahanukar A, Hallem EA, Carlson JR. Insect chemoreception. Current Opinion in Neurobiology. 2005;**15**(4):423-430

[18] Singh RN. Neurobiology of the gustatory systems of *Drosophila* and some terrestrial insects. Microscopy Research and Technique. 1997;**39**(6):547-563

[19] Schneider D. Insect Antennae. Annual Review of Entomology. 1964;**9**:103-122

[20] Larter NK, Sun J, Carlson JR. Organization and function of *Drosophila* odorant binding proteins. eLife. 2016;**5**:e20242

[21] Steinbrecht RA, Laue M, Maida R, Ziegelberger G. Odorant-binding proteins and their role in the detection of plant odours. In: Städler E, Rowell-Rahier M, Bauer R, editors. Proceedings of the 9th International Symposium on Insect-Plant Relationships. Dordrecht: Springer Netherlands; 1996. pp. 15-18

[22] Carlson JR. Olfaction in *Drosophila*: From odor to behavior. Trends in Genetics. 1996;**12**(5):175-180

[23] Pophof B, van der Goes van Naters W. Activation and inhibition of the transduction process in Silkmoth olfactory receptor neurons. Chemical Senses. 2002;**27**(5):435-443

[24] Nakagawa T, Cheng Y, Ramm E, Sheng M, Walz T. Structure and different conformational states of native AMPA receptor complexes. Nature. 2005;**433**(7025):545-549

[25] Sakurai T, Namiki S, Kanzaki R. Molecular and neural mechanisms of sex pheromone reception and processing in the silkmoth *Bombyx mori*. Frontiers in Physiology. 2014;**5**:125

[26] Sato K, Pellegrino M, Nakagawa T, Nakagawa T, Vosshall LB, Touhara K. Insect olfactory receptors are heteromeric ligand-gated ion channels. Nature. 2008;**452**(7190):1002-1006

[27] Wicher D, Schäfer R, Bauernfeind R, Stensmyr MC, Heller R, Heinemann SH, et al. *Drosophila* odorant receptors are both ligand-gated and cyclic-nucleotide-activated cation channels. Nature. 2008;**452**(7190):1007-1011

[28] Benton R, Vannice KS, Vosshall LB. An essential role for a CD36-related receptor in pheromone detection in *Drosophila*. Nature. 2007;**450**(7167):289-293

[29] Cassau S, Krieger J. The role of SNMPs in insect olfaction. Cell and Tissue Research. 2021;**383**(1):21-33

[30] Stengl M. Pheromone transduction in moths. Frontiers in Cellular Neuroscience. 2010;**4**:133

[31] Schoonhoven LM, van Loon J. An inventory of taste in caterpillars: Each species its own key. Acta Zoologica Academiae Scientiarum Hungaricae. 2002;**40**(Suppl. 1):215-263

[32] Shields VDC. Fine structure of the Galeal Styloconic Sensilla of larval *Lymantria dispar* (Lepidoptera: Lymantriidae). Annals of the Entomological Society of America. 2009;**102**(6):1116-1125

[33] Ishikawa S. Responses of maxillary chemoreceptors in the larva of the silkworm, *Bombyx mori*, to stimulation by carbohydrates. Journal of Cellular and Comparative Physiology. 1963;**61**:99-107

[34] Ishikawa S. Electrical response and function of a bitter substance receptor associated with the maxillary sensilla of the larva of the silkworm, *Bombyx*

mori L. Journal of Cellular Physiology. 1966;**67**(1):1-11

[35] Tsuneto K, Endo H, Shii F, Sasaki K, Nagata S, Sato R. Diet choice: The two-factor host acceptance system of silkworm larvae. PLOS Biology. 2020;**18**(9):e3000828

[36] Harrison GD. Host-plant discrimination and evolution of feeding preference in the *Colorado potato* beetle *Leptinotarsa decemlineata*. Physiological Entomology. 1987;**12**(4):407-415

[37] Blaney WM, Simmonds MSJ, Ley SV, Katz RB. An electrophysiological and behavioural study of insect antifeedant properties of natural and synthetic drimane-related compounds. Physiological Entomology. 1987;**12**(3):281-291

[38] Ma W. Dynamics of feeding responses in *Pieris brassicae* Linn. as a function of chemosensory input : A behavioural, ultrastructural and electrophysiological study [internal PhD, WU]. Wageningen: Veenman; 1972

[39] Glendinning JI, Davis A, Rai M. Temporal coding mediates discrimination of "bitter" taste stimuli by an insect. The Journal of Neuroscience. 2006;**26**(35):8900

[40] Glendinning JI, Davis A, Ramaswamy S. Contribution of different taste cells and signaling pathways to the discrimination of "bitter" taste stimuli by an insect. The Journal of Neuroscience. 2002;**22**(16):7281

[41] Sato K, Tanaka K, Touhara K. Sugar-regulated cation channel formed by an insect gustatory receptor. National Academy of Sciences of the United States of America. 2011;**108**(28):11680-11685

[42] Jiang X, Ning C, Guo H, Jia Y, Huang L, Qu M, et al. A gustatory receptor tuned to d-fructose in antennal sensilla chaetica of *Helicoverpa armigera*. Insect Biochemistry and Molecular Biology. 2015;**60**:39-46

[43] Vogt RG, Große-Wilde E, Zhou J. The Lepidoptera odorant binding protein gene family: Gene gain and loss within the GOBP/PBP complex of moths and butterflies. Insect Biochemistry and Molecular Biology. 2015;**62**:142-153

[44] Pelosi P, Mastrogiacomo R, Iovinella I, Tuccori E, Persaud KC. Structure and biotechnological applications of odorant-binding proteins. Applied Microbiology and Biotechnology. 2014;**98**(1):61-70

[45] Zhou J, Robertson G, He X, Dufour S, Hooper AM, Pickett JA, et al. Characterisation of *Bombyx mori* odorant-binding proteins reveals that a general odorant-binding protein discriminates between sex pheromone components. Journal of Molecular Biology. 2009;**389**(3):529-545

[46] Zhu J, Ban L, Song L-M, Liu Y, Pelosi P, Wang G. General odorant-binding proteins and sex pheromone guide larvae of *Plutella xylostella* to better food. Insect Biochemistry and Molecular Biology. 2016;**72**:10-19

[47] Zhu G, Zheng M, Sun J, Khuhro SA, Yan Q, Huang Y, et al. CRISPR/Cas9 mediated gene knockout reveals a more important role of PBP1 than PBP2 in the perception of female sex pheromone components in *Spodoptera litura*. Insect Biochemistry and Molecular Biology. 2019;**115**:103244

[48] Wang Q, Wang Q, Li H, Sun L, Zhang D, Zhang Y. Sensilla localization and sex pheromone recognition of odorant binding protein OBP4 in the mirid plant bug *Adelphocoris lineolatus*

(Goeze). Journal of Insect Physiology. 2020;**121**:104012

[49] Dahal SR, Lewellen JL, Ayyappan S, Chaudhary BP, Nukala V, Mohanty S. *Ostrinia furnacalis* PBP2 solution NMR structure: Insight into ligand binding and release mechanisms. Protein Science. 2022;**31**(10):e4438

[50] Hamiaux C, Carraher C, Löfstedt C, Corcoran JA. Crystal structure of *Epiphyas postvittana* pheromone binding protein 3. Scientific Reports. 2020;**10**(1):16366

[51] Zhang X, Mang D, Liao H, Ye J, Qian J, Dong S, et al. Functional disparity of three pheromone-binding proteins to different sex pheromone components in *Hyphantria cunea* (Drury). Journal of Agricultural and Food Chemistry. 2021;**69**(1):55-66

[52] Yasukochi Y, Yang B, Fujimoto T, Sahara K, Matsuo T, Ishikawa Y. Conservation and lineage-specific rearrangements in the GOBP/PBP gene complex of distantly related ditrysian Lepidoptera. PLoS One. 2018;**13**(2):e0192762

[53] Hu Y, Liu Y, Bi J, Chen Y, Zheng Y, Mao Y, et al. Field evaluation of sex pheromones and binding specificity of pheromone binding protein 4 in *Tryporyza intacta* (Lepidoptera: Crambidae). Scientific Reports. 2020;**10**(1):5464

[54] Vogt RG, Prestwich GD, Lerner MR. Odorant-binding-protein subfamilies associate with distinct classes of olfactory receptor neurons in insects. Journal of Neurobiology. 1991;**22**(1):74-84

[55] Zhou J, Zhang N, Wang P, Zhang S, Li D, Liu K, et al. Identification of host-plant volatiles and characterization of two novel general odorant-binding proteins from the legume pod borer, *Maruca vitrata* Fabricius (Lepidoptera: Crambidae). PLoS One. 2015;**10**(10):e0141208

[56] Liu N, Yang K, Liu Y, Xu W, Anderson A, Dong S. Two general-odorant binding proteins in *Spodoptera litura* are differentially tuned to sex pheromones and plant odorants. Comparative Biochemistry and Physiology Part A: Molecular & Integrative Physiology. 2015;**180**:23-31

[57] Chen X, Li B, Chen Y, Li G, Wu J. Functional analysis of the odorant receptor coreceptor in odor detection in *Grapholita molesta* (Lepidoptera: Tortricidae). Archives of Insect Biochemistry and Physiology. 2021;**108**(2):e21837

[58] Benton R. On the ORigin of smell: Odorant receptors in insects. Cellular and Molecular Life Sciences CMLS. 2006;**63**(14):1579-1585

[59] Fleischer J, Pregitzer P, Breer H, Krieger J. Access to the odor world: Olfactory receptors and their role for signal transduction in insects. Cellular and Molecular Life Sciences : CMLS. 2018;**75**(3):485-508

[60] Robertson HM. Molecular evolution of the major arthropod chemoreceptor gene families. Annual Review of Entomology. 2019;**64**:227-242

[61] Fan X, Mo B, Li G, Huang L, Guo H, Gong X, et al. Mutagenesis of the odorant receptor co-receptor (Orco) reveals severe olfactory defects in the crop pest moth *Helicoverpa armigera*. BMC Biology. 2022;**20**(1):214

[62] Sun H, Bu L, Su S, Guo D, Gao C, Wu S. Knockout of the odorant receptor co-receptor, orco, impairs feeding, mating and egg-laying behavior in the

fall armyworm *Spodoptera frugiperda*. Insect Biochemistry and Molecular Biology. 2023;**152**:103889

[63] Fleischer J, Krieger J. Insect pheromone receptors – Key elements in sensing intraspecific chemical signals. Frontiers in Cellular Neuroscience. 2018;**12**:425

[64] Yuvaraj JK, Corcoran JA, Andersson MN, Newcomb RD, Anderbrant O, Löfstedt C. Characterization of odorant receptors from a non-ditrysian moth, *Eriocrania semipurpurella* sheds light on the origin of sex pheromone receptors in Lepidoptera. Molecular Biology and Evolution. 2017;**34**(11):2733-2746

[65] Cao S, Huang T, Shen J, Liu Y, Wang G. An orphan pheromone receptor affects the mating behavior of *Helicoverpa armigera*. Frontiers in Physiology. 2020;**11**:413

[66] Koutroumpa F, Monsempès C, Anton S, François M-C, Montagné N, Jacquin-Joly E. Pheromone receptor knock-out affects pheromone detection and brain structure in a moth. Biomolecules. 2022;**12**(3):341

[67] Benton R, Vannice KS, Gomez-Diaz C, Vosshall LB. Variant ionotropic glutamate receptors as chemosensory receptors in *Drosophila*. Cell. 2009;**136**(1):149-162

[68] Zhang J, Bisch-Knaden S, Fandino RA, Yan S, Obiero GF, Grosse-Wilde E, et al. The olfactory coreceptor IR8a governs larval feces-mediated competition avoidance in a hawkmoth. National Academy of Sciences of the United States of America. 2019;**116**(43):21828-21833

[69] Zhang X, Yang B, Sun D, Guo M, Zhang J, Wang G. Ionotropic receptor 8a is involved in the attraction of *Helicoverpa armigera* to acetic acid. Insect Science. 2022;**29**(3):657-668

[70] Hou X, Zhang D, Powell D, Wang H, Andersson MN, Löfstedt C. Ionotropic receptors in the turnip moth Agrotis segetum respond to repellent medium-chain fatty acids. BMC Biology. 2022;**20**(1):34

[71] Cui Z, Gao Z, Zhang W, Li Y, Sun J, Mang D. Glucose-Induce Activation of CCHamide2 via Gustatory Receptor to Regulate Metabolism in Hyphantria cunea Larvae. China: College of Life Science, Hebei University, Baoding. Unpublished manuscript

[72] McKenna MP, Hekmat-Scafe DS, Gaines P, Carlson JR. Putative *Drosophila* pheromone-binding proteins expressed in a subregion of the olfactory system. Journal of Biological Chemistry. 1994;**269**(23):16340-16347

[73] Yang H, Dong J, Sun Y, Hu Z, Lv Q, Li D. Antennal transcriptome analysis and expression profiles of putative chemosensory soluble proteins in *Histia rhodope* Cramer (Lepidoptera: Zygaenidae). Comparative Biochemistry and Physiology Part D: Genomics and Proteomics. 2020;**33**:100654

[74] Duan S, Li D, Wang M. Chemosensory proteins used as target for screening behaviourally active compounds in the rice pest *Cnaphalocrocis medinalis* (Lepidoptera: Pyralidae). Insect Molecular Biology. 2019;**28**(1):123-135

[75] Pelosi P, Iovinella I, Zhu J, Wang G, Dani FR. Beyond chemoreception: Diverse tasks of soluble olfactory proteins in insects. Biological Reviews. 2018;**93**(1):184-200

[76] Zhang Y, Ye Z, Yang K, Dong S. Antenna-predominant and male-biased

CSP19 of *Sesamia inferens* is able to bind the female sex pheromones and host plant volatiles. Gene. 2014;**536**(2):279-286

[77] Li Z, Zhang S, Luo J, Zhu J, Cui J, Dong S. Expression analysis and binding assays in the chemosensory protein gene family indicate multiple roles in *Helicoverpa armigera*. Journal of Chemical Ecology. 2015;**41**(5):473-485

[78] Xuan N, Guo X, Xie H, Lou Q, Lu X, Liu G, et al. Increased expression of CSP and CYP genes in adult silkworm females exposed to avermectins. Insect Science. 2015;**22**(2):203-219

[79] Yao Q, Liang Z, Chen B. Evidence for the participation of chemosensory proteins in response to insecticide challenge in *Conopomorpha sinensis*. Journal of Agricultural and Food Chemistry. 2023;**71**(3):1360-1368

[80] Vogt RG, Riddiford LM. Pheromone binding and inactivation by moth antennae. Nature. 1981;**293**(5828):161-163

[81] He P, Li Z, Liu C, Liu S, Dong S. Two esterases from the genus *Spodoptera* degrade sex pheromones and plant volatiles. Genome. 2014;**57**:1-8

[82] He P, Zhang Y, Li Z, Yang K, Zhu J, Liu S, et al. An antennae-enriched carboxylesterase from *Podoptera exigua* displays degradation activity in both plant volatiles and female sex pheromones. Insect Molecular Biology. 2014;**23**(4):475-486

[83] Wang M, Long G, Guo H, Liu X, Wang H, Dewer Y, et al. Two carboxylesterase genes in *Plutella xylostella* associated with sex pheromones and plant volatiles degradation. Pest Management Science. 2021;**77**(6):2737-2746

[84] He P, Zhang Y, Hong D, Wang J, Wang X, Zuo L, et al. A reference gene set for sex pheromone biosynthesis and degradation genes from the diamondback moth, *Plutella xylostella*, based on genome and transcriptome digital gene expression analyses. BMC Genomics. 2017;**18**(1):219

[85] Wang M, He M, Wang H, Ma Y, Dewer Y, Zhang F, et al. A candidate aldehyde oxidase in the antennae of the diamondback moth, *Plutella xylostella* (L.), is potentially involved in the degradation of pheromones, plant-derived volatiles and the detoxification of xenobiotics. Pesticide Biochemistry and Physiology. 2021;**171**:104726

[86] Wei H, Tan S, Li Z, Li J, Moural TW, Zhu F, et al. Odorant degrading carboxylesterases modulate foraging and mating behaviors of *Grapholita molesta*. Chemosphere. 2021;**270**:128647

[87] Garattini E, Fratelli M, Terao M. Mammalian aldehyde oxidases: Genetics, evolution and biochemistry. Cellular and Molecular Life Sciences. 2008;**65**(7):1019-1048

[88] Godoy R, Arias I, Venthur H, Quiroz A, Mutis A. Characterization of two aldehyde oxidases from the greater wax moth, *Galleria mellonella* Linnaeus. (Lepidoptera: Pyralidae) with potential role as odorant-degrading enzymes. Insects [Internet]. 2022;**13**(12)

[89] Tijet N, Helvig C, Feyereisen R. The cytochrome P450 gene superfamily in *Drosophila melanogaster*: Annotation, intron-exon organization and phylogeny. Gene. 2001;**262**(1):189-198

[90] Younus F, Fraser NJ, Coppin CW, Liu J, Correy GJ, Chertemps T, et al. Molecular basis for the behavioral effects of the odorant degrading enzyme

esterase 6 in *Drosophila*. Scientific Reports. 2017;7(1):46188

[91] Chertemps T, Maïbèche M. 19 - Odor degrading enzymes and signal termination. In: Blomquist GJ, Vogt RG, editors. Insect Pheromone Biochemistry and Molecular Biology. 2nd ed. London: Academic Press; 2021. pp. 619-644

[92] Vogt RG, Prestwich GD, Riddiford LM. Sex pheromone receptor proteins. Visualization using a radiolabeled photoaffinity analog. Journal of Biological Chemistry. 1988;**263**(8):3952-3959

[93] Rogers ME, Sun M, Lerner MR, Vogt RG. Snmp-1, a novel membrane protein of olfactory neurons of the silk moth *Antheraea polyphemus* with homology to the CD36 family of membrane proteins. Journal of Biological Chemistry. 1997;**272**(23):14792-14799

[94] Rogers ME, Steinbrecht RA, Vogt RG. Expression of SNMP-1 in olfactory neurons and sensilla of male and female antennae of the silkmoth *Antheraea polyphemus*. Cell and Tissue Research. 2001;**303**(3):433-446

[95] Xu W, Zhang H, Liao Y, Papanicolaou A. Characterization of sensory neuron membrane proteins (SNMPs) in cotton bollworm *Helicoverpa armigera* (Lepidoptera: Noctuidae). Insect Science. 2021;**28**(3):769-779

[96] Liu X, Sun S, Hou W, Zhang J, Yan Q, Dong S. Functional characterization of two spliced variants of fructose gustatory receptor in the diamondback moth, *Plutella xylostella*. Pesticide Biochemistry and Physiology. 2020;**164**:7-13

[97] Zhang H, Xu W, Chen Q, Sun L, Anderson A, Xia Q, et al. A phylogenomics approach to characterizing sensory neuron membrane proteins (SNMPs) in Lepidoptera. Insect Biochemistry and Molecular Biology. 2020;**118**:103313

[98] Dahanukar A, Foster K, van der Goes van Naters WM, Carlson JR. A Gr receptor is required for response to the sugar trehalose in taste neurons of Drosophila. Nature Neuroscience. 2001;**4**(12):1182-1186

[99] Dahanukar A, Lei Y, Kwon JY, Carlson JR. Two Gr genes underlie sugar reception in *Drosophila*. Neuron. 2007;**56**(3):503-516

[100] Jiao Y, Moon SJ, Wang X, Ren Q, Montell C. Gr64f is required in combination with other gustatory receptors for sugar detection in *Drosophila*. Current Biology. 2008;**18**(22):1797-1801

[101] Fujii S, Yavuz A, Slone J, Jagge C, Song X, Amrein H. *Drosophila* sugar receptors in sweet taste perception, olfaction, and internal nutrient sensing. Current Biology. 2015;**25**(5):621-627

[102] Mang D, Mayu K, Toyama T, Yamagishi T, Sato R. BmGr4 responds to sucrose and glucose and expresses in tachykinin-related peptide-secreting enteroendocrine cells. Insect Biochemistry and Molecular Biology. 2022;**150**:103858

[103] Mang D, Shu M, Endo H, Yoshizawa Y, Nagata S, Kikuta S, et al. Expression of a sugar clade gustatory receptor, BmGr6, in the oral sensory organs, midgut, and central nervous system of larvae of the silkworm *Bombyx mori*. Insect Biochemistry and Molecular Biology. 2016;**70**:85-98

[104] Mang D, Toyama T, Yamagishi T, Sun J, Purba ER, Endo H, et al. Dietary compounds activate an insect gustatory receptor on enteroendocrine cells to

elicit myosuppressin secretion. Insect Biochemistry and Molecular Biology. 2023;**155**:103927

[105] Zhang S, Wang P, Ning C, Yang K, Li G, Cao L, et al. The larva and adult of *Helicoverpa armigera* use differential gustatory receptors to sense sucrose. eLife. 2024;**12**:RP91711

[106] Miyamoto T, Slone J, Song X, Amrein H. A fructose receptor functions as a nutrient sensor in the Drosophila brain. Cell. 2012;**151**(5):1113-1125

[107] Wanner KW, Robertson HM. The gustatory receptor family in the silkworm moth *Bombyx mori* is characterized by a large expansion of a single lineage of putative bitter receptors. Insect Molecular Biology. 2008;**17**(6):621-629

[108] Ai D, Dong C, Yang B, Yu C, Wang G. A fructose receptor gene influences development and feed intake in *Helicoverpa armigera*. Insect Science. 2022;**29**(4):993-1005

[109] Liu X, Yan Q, Yang Y, Hou W, Miao C, Peng Y, et al. A gustatory receptor GR8 tunes specifically to D-fructose in the common cutworm *Spodoptera litura*. Insects [Internet]. 2019;**10**(9)

[110] Sun J, Zhang W, Cui Z, Pan Y, Smagghe G, Zhang L, et al. HcGr76 responds to fructose and chlorogenic acid and is involved in regulation of peptide expression in the midgut of larvae. Pest Management Science. 2024;**80**(11):5672-5683

[111] Zhang H, Anderson AR, Trowell SC, Luo A, Xiang Z, Xia Q. Topological and functional characterization of an insect gustatory receptor. PLoS One. 2011;**6**(8):e24111

[112] Kikuta S, Endo H, Tomita N, Takada T, Morita C, Asaoka K, et al. Characterization of a ligand-gated cation channel based on an inositol receptor in the silkworm, *Bombyx mori*. Insect Biochemistry and Molecular Biology. 2016;**74**:12-20

[113] Endo H, Tsuneto K, Mang D, Zhang W, Yamagishi T, Ito K, et al. Molecular basis of host plant recognition by silkworm larvae. Journal of Insect Physiology. 2024;**154**:104628

[114] Moon SJ, Lee Y, Jiao Y, Montell C. A *Drosophila* gustatory receptor essential for aversive taste and inhibiting male-to-male courtship. Current Biology. 2009;**19**(19):1623-1627

[115] Lee Y, Kim SH, Montell C. Avoiding DEET through insect gustatory receptors. Neuron. 2010;**67**(4):555-561

[116] Poudel S, Kim Y, Gwak J, Jeong S, Lee Y. Gustatory receptor 22e is essential for sensing chloroquine and strychnine in *Drosophila melanogaster*. Insect Biochemistry and Molecular Biology. 2017;**88**:30-36

[117] Lee Y, Moon SJ, Wang Y, Montell C. A *Drosophila* gustatory receptor required for strychnine sensation. Chemical Senses. 2015;**40**(7):525-533

[118] Shim J, Lee Y, Jeong YT, Kim Y, Lee MG, Montell C, et al. The full repertoire of *Drosophila* gustatory receptors for detecting an aversive compound. Nature Communications. 2015;**6**(1):8867

[119] Moon SJ, Köttgen M, Jiao Y, Xu H, Montell C. A taste receptor required for the caffeine response in vivo. Current Biology. 2006;**16**(18):1812-1817

[120] Lee Y, Moon SJ, Montell C. Multiple gustatory receptors required for the caffeine response in *Drosophila*. National Academy of Sciences

of the United States of America. 2009;**106**(11):4495-4500

[121] Poudel S, Lee Y. Gustatory receptors required for avoiding the toxic compound Coumarin in *Drosophila melanogaster*. Molecules and Cells. 2016;**39**(4):310-315

[122] Sang J, Rimal S, Lee Y. Gustatory receptor 28b is necessary for avoiding saponin in *Drosophila melanogaster*. EMBO Reports. 2019;**20**(2):e47328

[123] Ozaki K, Ryuda M, Yamada A, Utoguchi A, Ishimoto H, Calas D, et al. A gustatory receptor involved in host plant recognition for oviposition of a swallowtail butterfly. Nature Communications. 2011;**2**(1):542

[124] Zhang Z, Zhang S, Niu B, Ji D, Liu X, Li M, et al. A determining factor for insect feeding preference in the silkworm, *Bombyx mori*. PLOS Biology. 2019;**17**(2):e3000162

[125] Kasubuchi M, Shii F, Tsuneto K, Yamagishi T, Adegawa S, Endo H, et al. Insect taste receptors relevant to host identification by recognition of secondary metabolite patterns of non-host plants. Biochemical and Biophysical Research Communications. 2018;**499**(4):901-906

[126] Zhang S, Tang J, Li Y, Li D, Chen G, Chen L, et al. The silkworm gustatory receptor BmGr63 is dedicated to the detection of isoquercetin in mulberry. Proceedings of the Royal Society B: Biological Sciences. 1985;**2022**(289):20221427

[127] Zhang W, Lv D, Mang D. BmGr67 is a Functional Gustatory Receptor to Response Berberine. China: College of Life Science, Hebei University, Baoding; Unpublished manuscript

[128] Yang K, Gong X, Li G, Huang L, Ning C, Wang C. A gustatory receptor tuned to the steroid plant hormone brassinolide in *Plutella xylostella* (Lepidoptera: Plutellidae). eLife. 2020;**9**:e64114

[129] Yang J, Guo H, Jiang N, Tang R, Li G, Huang L, et al. Identification of a gustatory receptor tuned to sinigrin in the cabbage butterfly *Pieris rapae*. PLOS Genetics. 2021;**17**(7):e1009527

[130] Chen Y, Wang P, Zhang S, Yang J, Li G, Huang L, et al. Functional analysis of a bitter gustatory receptor highly expressed in the larval maxillary galea of *Helicoverpa armigera*. PLOS Genetics. 2022;**18**(10):e1010455

[131] Zhang W, Liu H, Mang D. Molecular Basis of Quinine-Induced Feeding Suppression in Hyphantria cunea Larvae. China: College of Life Science, Hebei University, Baoding. Unpublished manuscript

[132] Guo H, Cheng T, Chen Z, Jiang L, Guo Y, Liu J, et al. Expression map of a complete set of gustatory receptor genes in chemosensory organs of *Bombyx mori*. Insect Biochemistry and Molecular Biology. 2017;**82**:74-82

[133] Briscoe AD, Macias-Muñoz A, Kozak KM, Walters JR, Yuan F, Jamie GA, et al. Female behaviour drives expression and evolution of gustatory receptors in butterflies. PLOS Genetics. 2013;**9**(7):e1003620

[134] Engsontia P, Sangket U, Chotigeat W, Satasook C. Molecular evolution of the odorant and gustatory receptor genes in lepidopteran insects: Implications for their adaptation and speciation. Journal of Molecular Evolution. 2014;**79**(1):21-39

[135] Suzuki HC, Ozaki K, Makino T, Uchiyama H, Yajima S, Kawata M. Evolution of gustatory receptor gene

family provides insights into adaptation to diverse host plants in nymphalid butterflies. Genome Biology and Evolution. 2018;**10**(6):1351-1362

[136] Rasch C, Rembold H. Carbon-dioxide — Highly attractive signal for larvae of *Helicoverpa armigera*. Die Naturwissenschaften. 1994;**81**(5):228-229

[137] Huang X, Mack TP. Artificial carbon dioxide source to attract lesser cornstalk borer (Lepidoptera: Pyralidae) larvae. Journal of Economic Entomology. 2001;**94**(4):860-867

[138] Hew CS, Thio YC, Wong SY, Chin TY. Rhythmic production of CO_2 by tropical orchid flowers. Physiologia Plantarum. 1978;**42**(2):226-230

[139] Seymour RS, White CR, Gibernau M. Heat reward for insect pollinators. Nature. 2003;**426**(6964):243-244

[140] Stange G, Monro J, Stowe S, Osmond CB. The CO_2 sense of the moth *Cactoblastis cactorum* and its probable role in the biological control of the CAM plant *Opuntia stricta*. Oecologia. 1995;**102**(3):341-352

[141] Guerenstein PG, Hildebrand JG. Roles and effects of environmental carbon dioxide in insect life. Annual Review of Entomology. 2008;**53**:161-178

[142] Stange G. Effects of changes in atmospheric carbon dioxide on the location of hosts by the moth, *Cactoblastis cactorum*. Oecologia. 1997;**110**(4):539-545

[143] Abrell L, Guerenstein PG, Mechaber WL, Stange G, Christensen TA, Nakanishi K, et al. Effect of elevated atmospheric CO_2 on oviposition behavior in *Manduca sexta* moths. Global Change Biology. 2005;**11**(8):1272-1282

[144] Choi KS, Ahn S-J, Kim SB, Ahn JJ, Jung BN, Go SW, et al. Elevated CO_2 may alter pheromonal communication in *Helicoverpa armigera* (Lepidoptera: Noctuidae). Physiological Entomology. 2018;**43**(3):169-179

[145] Jones WD, Cayirlioglu P, Grunwald Kadow I, Vosshall LB. Two chemosensory receptors together mediate carbon dioxide detection in *Drosophila*. Nature. 2007;**445**(7123):86-90

[146] Spaethe J, Briscoe AD. Early duplication and functional diversification of the opsin gene family in insects. Molecular Biology and Evolution. 2004;**21**(8):1583-1594

[147] Liu Y, Guo H, Huang L, Pelosi P, Wang C. Unique function of a chemosensory protein in the proboscis of two *Helicoverpa species*. Journal of Experimental Biology. 2014;**217**(10):1821-1826

[148] Xu W, Anderson A. Carbon dioxide receptor genes in cotton bollworm *Helicoverpa armigera*. The Science of Nature. 2015;**102**(3):11

[149] Briscoe AD. Six opsins from the butterfly *Papilio glaucus*: Molecular phylogenetic evidence for paralogous origins of red-sensitive visual pigments in insects. Journal of Molecular Evolution. 2000;**51**(2):110-121

[150] Liu Z, Wang X, Lei C, Zhu F. Sensory genes identification with head transcriptome of the migratory armyworm, *Mythimna separata*. Scientific Reports. 2017;**7**(1):46033

[151] Zhang J, Wang B, Dong S, Cao D, Dong J, Walker WB, et al. Antennal transcriptome analysis and comparison

of chemosensory gene families in two closely related Noctuidae Moths, *Helicoverpa armigera* and H. Assulta. PLoS One. 2015;**10**(2):e0117054

[152] Kanost MR, Arrese EL, Cao X, Chen Y, Chellapilla S, Goldsmith MR, et al. Multifaceted biological insights from a draft genome sequence of the tobacco hornworm moth, *Manduca sexta*. Insect Biochemistry and Molecular Biology. 2016;**76**:118-147

[153] Ning C, Yang K, Xu M, Huang L, Wang C. Functional validation of the carbon dioxide receptor in labial palps of *Helicoverpa armigera* moths. Insect Biochemistry and Molecular Biology. 2016;**73**:12-19

[154] Arce CCM, Theepan V, Schimmel BCJ, Jaffuel G, Erb M, Machado RAR. Plant-associated CO_2 mediates long-distance host location and foraging behaviour of a root herbivore. eLife. 2021;**10**:e65575

[155] Jiang X, Guo H, Di C, Yu S, Zhu L, Huang L, et al. Sequence similarity and functional comparisons of pheromone receptor orthologs in two closely related *Helicoverpa species*. Insect Biochemistry and Molecular Biology. 2014;**48**:63-74

[156] Kurtovic A, Widmer A, Dickson BJ. A single class of olfactory neurons mediates behavioural responses to a *Drosophila* sex pheromone. Nature. 2007;**446**(7135):542-546

[157] Freeman EG, Wisotsky Z, Dahanukar A. Detection of sweet tastants by a conserved group of insect gustatory receptors. National Academy of Sciences of the United States of America. 2014;**111**(4):1598-1603

[158] Guo H, Huang L, Gong X, Wang C. Comparison of functions of pheromone receptor repertoires in *Helicoverpa armigera* and Helicoverpa assulta using a *Drosophila* expression system. Insect Biochemistry and Molecular Biology. 2022;**141**:103702

[159] Wang L, Anderson DJ. Identification of an aggression-promoting pheromone and its receptor neurons in *Drosophila*. Nature. 2010;**463**(7278):227-231

[160] Hou X, Zhang D, Yuvaraj JK, Corcoran JA, Andersson MN, Löfstedt C. Functional characterization of odorant receptors from the moth *Eriocrania semipurpurella*: A comparison of results in the Xenopus oocyte and HEK cell systems. Insect Biochemistry and Molecular Biology. 2020;**117**:103289

[161] Pellegrino M, Nakagawa T, Vosshall LB. Single Sensillum Recordings in the Insects *Drosophila Melanogaster* and *Anopheles gambiae*. Journal of Visualized Experiments. 2010;(36):1-5

[162] Halty-deLeon L, Pal Mahadevan V, Wiesel E, Hansson BS, Wicher D. Response plasticity of *Drosophila* olfactory sensory neurons. International Journal of Molecular Sciences [Internet]. 2024;**25**(13)

Chapter 3

Spraysable Sex Pheromones for Codling Moth: Waiting for Grower Adoption

Alan Lee Knight, Esteban Basoalto and Michele Preti

Abstract

Microencapsulated (MEC) formulations for mating disruption (MD) for codling moth, *Cydia pomonella* (L.), were developed to cover the entire crop canopy using air blast sprayers, saving growers $$ from the increasing labor costs applying dispensers. Treating orchards with MEC sprays only when pest densities exceed thresholds provide growers greater flexibility than dispensers. However, early research suggested that MECs did not disrupt male catch as well as dispensers and were less effective under higher pest pressures. The development of low volume MEC applications proved this was incorrect. A new MEC formulation with much smaller microcapsules was registered in 2019 and provides greater rainfastness and longevity. Dissections of female codling moths shows the new MEC performs as well or better than hand-applied dispensers. Combined MEC + dispenser programs further improved the effectiveness of codling moth MD.

Keywords: microencapsulation, *Cydia pomonella*, *Grapholita molesta*, pear ester, mating disruption

1. Introduction

The era of sex pheromone-based mating disruption (MD) for agricultural pests began nearly 60 years ago [1]. Researchers quickly considered the advantages in treating large fields with sprayable materials [2]. Unfortunately, specialized equipment was necessary to apply these formulations (hollow fibers and laminated flakes), and this slowed early adoption of MD [3, 4]. The same issue continued to limit adoption of sprayable materials (wax-based drops) later even while hand-applied MD technologies were being widely adopted [5, 6].

The development of microencapsulated (MEC) formulations compatible with standard air blast sprayers and in tank mixes with other materials was seen as a promising approach [7]. Microcapsules can be created via several processes with coacervation-phase separation, the most common [8, 9]. The formulation goal is to keep capsules intact and to have a flat first-order release over an extended period [10]. Chemists can adjust a suite of factors to create capsules with different properties, and these are generally kept confidential [11]. The first use of microencapsulation of a sex pheromone was reported with gypsy moth, *Lymantria dispar* (L.) (Lepidoptera: Erebidae), [12]. Only a few companies have pursued the development of microencapsulated formulations. Consep Membrane Inc. (Suterra Inc.) initiated CheckMate® flowables in 1997 [13], and 3 M developed products beginning in 1999 [14]. 3 M discontinued their support for this technology in 2006. Trécé Inc. developed a new MEC technology (CIDETRAK® MEC™) in 2016. Future formulations may include the use of nanoparticles [15].

2. Background

Commercial development of MEC formulations has primarily included two tortricid pests of tree fruits, codling moth (CM), *Cydia pomonella* (L.) and oriental fruit moth (OFM), *Grapholita molesta* (Busck) (Lepidoptera: Tortricidae). Mating disruption is a widely used tactic to manage these pests [6]. Only a few studies with MEC products have been reported with other tortricids: *Sparganothis sulfureana* Clemens, in blueberries [16], *Eupoecilia ambiguella* (Hübner) and *Paralobesia viteana* (Clemens) in grapes [17–19], and *Argyrotaenia velutinana* (Walker), and *Choristoneura rosaceana* (Harris) in pome fruits [20, 21]. Our review of published studies with CM includes both peer reviewed journal articles and various incomplete reports from scientists/grower meetings or to commodity commissions.

All CM MEC products are loaded with codlemone (*E,E*)-8,10-dodecadien-1-ol [22]. The CheckMate® CM-F product is characterized as microbeads with 50–150 μm diameters [23]. The CheckMate® CM-F product is a 14.4% formulation and is registered at 10–20 g a.i. ha^{-1} per application. The 3 M-CM product was a 9% formulation applied at 15–25 g a.i. ha^{-1}. The capsule sizes of the various 3 M MEC products were never published. Trécé Inc. registered a sex pheromone formulation for CM MD (CIDETRAK® CM MEC™) and a formulation with pear ester, (2*E*,4*Z*)-ethyl-2,4-decadienoate (CIDETRAK® DA MEC™) to enhance both CM larval and adult management [24, 25]. The CIDETRAK® CM MEC™ is a 7% formulation used at 17.5–50 g a.i. ha^{-1} per application and limited to a total seasonal use of 370 g a.i. ha^{-1}. The pear ester formulation is characterized as 25.9×10^4 microcapsules per mL with 2–14 μm capsule diameters [26]. The CIDETRAK® DA MEC™ is a 5% formulation registered for use at 1.5–8.75 g a.i. ha^{-1} per application and not exceeding 8 applications and 12 g a.i. ha^{-1} per year.

2.1 Factors affecting the efficacy of MEC formulations

The goal of MEC applications is to deposit and retain as many microcapsules for as long as possible throughout the canopy. This is somewhat different than the typical complete coverage required with most pesticides. Thus, MEC sprays applied with traditional air blast sprayers can be targeted to specific locations within tree canopies by adjusting and turning off some nozzles. Lower spray volumes using custom sprayers have also been used to reduce labor costs.

2.1.1 Deposition, distribution, and retention

The clever use of a fluorescent dye made it possible to easily count the CheckMate® CM-F microbeads on leaves and examine differences between and among leaf surfaces within the canopy [27–29]. Similarly, staining of plant materials was used to enhance the counting of the 3 M capsules [21, 30]. The impact of leaf surfaces, cultivars, and host crops were found to be significant factors impacting deposition [31]. The distribution of microcapsules was mapped, and the highest numbers were on the underside of leaves in the direct spray zone [27]. High levels of capsules were also deposited on the upper surface of leaves below the spray zone and on leaves in the canopy of trees in the next row. Low volume MEC applications with slower air speed increased the deposition of the CheckMate® products [28, 32]. Similar results were found for the 3 M-OFM microcapsules in stone fruit [30].

Retention of the microcapsules is another key concern with sprayable technology. Rainfall, overhead irrigation, and evaporative cooling used to reduce sunburn in apples are each important factors impacting retention. In the arid, western tree fruit production areas of USA summer rain can be infrequent for several months and rain events are typically <1.5 cm. Interest in sprayable MD technology is likely lower in production regions characterized by greater and more frequent rainfall due to wash-off. The use of overhead irrigation in tree fruit today is rare having been replaced by more efficient under tree drip irrigation. However, the use of overhead water with micro sprinklers to cool the canopy in the later third of the growing season is common in the arid production regions [33]. Laboratory studies with simulated rainfall found an inverse relationship with the amount of water applied [32, 34]. Sloping leaves retained fewer microcapsules [23]. Overhead irrigation reduced the efficacy of the CheckMate® CM-F product in orchards [35]. Orchards with overhead versus under tree irrigation had nearly 3-fold higher moth catches and 2-fold higher levels of fruit injury (**Table 1**) [37].

To improve the deposition and retention of microcapsules most studies have included the use of stickers which have increased both deposition and retention [23, 39]. Different results have been reported. The use of the pine resin product (Nu-Film, Miller Chemical, Hanover, PA) improved the two OFM formulations but not the CM sprayable products [21]. Similarly, the use of a latex sticker did not increase the capsule deposition of a low-volume CM-F application [37].

2.1.2 Longevity

MEC capsules deposited in the orchard are exposed to variable temperatures and UV radiation during the season [40]. Pheromone emissions from capsules generally have a short burst followed by an extended first-order rate [11]. A portable electroantennogram was used to measure the longevity of 3 M-CM capsules in the field of only two weeks [41]. CheckMate® CM-F capsules remained effective under shade conditions and retained some activity even under direct sunlight for up to four weeks [42]. Particle films, such as kaolin clay, used for sunburn protection reduced the emission rates of CheckMate® flowable sprays by 22% [43]. Unfortunately, no study considered whether particle films could be used to either extend the retention of microcapsules or their longevity. However, the use of a 3 M sprayable formulation for the leafroller *C. rosaceana* combined with 2% horticultural oil slowed its emission rate and extended its effectiveness in disrupting virgin female-baited traps in both laboratory and field studies [44, 45]. Another approach to extend MEC effectiveness has been to apply lower rates of MEC more frequently [29].

2.1.3 Comparison of formulations

Evaluations of MEC formulations as well as other MD technologies have typically been measured by reductions in male catch in sex pheromone-baited traps and

Year	MEC	Rate per spray g a.i. ha^{-1}	No. applications/spray interval/ spray volume ha^{-1}		Reduction compared to untreated (UTC), insecticide-only, or MD dispensers		References
					Male catch in PH (VF)- baited traps	Level of injury	
1999	3 M-CM	100	5/28–35 d/935 L	**HIGH**	60–80%	—	[17]
2000	CM-F	49	1/29 d/46 L	**LOW**	(90%) WK 1 and (75%) WK 3 < UTC	—	[32]
2000	CM-F	49	1/29 d/926 L	**HIGH**	(55%) WK 1 and (0%) WK 3 < UTC	—	
2004	CM-F	25	4/30 d/935 L	**HIGH**	93% < UTC	46% < insecticide	[36]
2004	3 M-CM	25	4/30 d/935 L	**HIGH**	95% < UTC	29% < insecticide	
2004	CM-F	25	6/21–28 d/12–23 L	**LOW**	80–90% < UTC	85–88% < UTC	[37]
2004	CM-F	25	6/21–28 d/935 L	**HIGH**	61% < UTC	69% < UTC	
2005	CM-F	25	4/24–28 d/46 L	**LOW**	3X > dispensers	= dispensers	[37]
2005	CM-F	25	5/21–28 d/46 L	**LOW**	40% > dispensers	= dispensers	
2006–2007	CM-F	25	2/4 wk./24 L	**LOW**	77–82% < UTC	-	[28]
2006–2007	CM-F	25	2/4 wk./768 L	**HIGH**	40–50% < UTC	-	
2009–2010	CM-F + DAMEC	25 + 5	4/2 wk./40 L	**LOW**	82–88% < UTC	87–92% < UTC	[38]
	CM-F + DAMEC	25 + 5	4/2 wk./635 L	**HIGH**	82–87% < UTC	76–88% < UTC	

Table 1.
Summary of field evaluations of 3 M-CM or CheckMate® CM-F formulations for codling moth (CM), Cydia pomonella, *using different rate (25–100 g a.i.* ha^{-1}*) and spray volumes (***LOW** *12–46 L* ha^{-1} *or* **HIGH** *635–1000 L* ha^{-1}*) comparing male moth catch in traps and levels of fruit injury, 1999–2010.*

levels of fruit/nut injury, either compared with an untreated control, a standard insecticide program, or against dispensers. **Table 1** summarizes these trials of 3 M and CheckMate® products for CM in nine trials conducted from 1999 to 2006. The significance of data from several more incomplete data sets from reports will also be included in this comparison.

The earliest report with MEC used 3 M-CM at a high rate (100 g a.i. ha^{-1}) (**Table 1**) [17]. This rate was cut in half in the first study with CheckMate® CM-F [32]. Subsequent studies with CM looked at further rate reductions and the standard rate became 25 g a.i. ha^{-1} [28, 36]. MEC programs either used one spray per generation or regular schedules of sprays spaced four weeks apart. However, both disruption and fruit injury were further improved with a 9-spray program using lower rates and less material per season compared with a standard 2-spray program [21].

A number of similar studies were conducted with MEC formulations from both companies for OFM. The 3 M-OFM formulation applied in three sprays was shown to be effective and performed similarly to the use of dispensers in North Carolina and Canada [46, 47]. Five or six sprays of a higher rate of 3 M-OFM in peach and pear disrupted traps and protected fruit as well as dispensers in Australia [48]. The 3 M-OFM was used late in the season to extend the effectiveness of a spring dispenser application [47]. Only three studies evaluated the OFM-F product [21, 49, 50]. The two OFM formulations performed well and similarly when applied at a standard spray volume or as alternate-row-middle half-sprays [21, 49]. The OFM-F outperformed the use of OFM dispensers when applied in only 12 L ha^{-1} [50].

In general, the MEC formulations for CM were not as effective as concurrent OFM trials in Michigan and Pennsylvania apple [21, 49]. Unfortunately fruit injury from CM and OFM was not separated in the latter study. Studies evaluated low rates and frequent applications (2.5–5.0 g a.i. every 10–14 d) versus less frequent and higher rates (20–30 g a.i. once per generation) for both species using 3 M products with a standard spray volume [39]. The repeated low-rate applications improved the management of OFM but not for CM. Similarly, nine low-rate applications were more effective in reducing fruit injury than the use of higher rates applied once per generation in a combined program for OFM and CM [21].

CM is an important pest of walnuts *Juglans regia* L., in California, but their large size makes the use of hand-applied dispensers difficult [51]. The use of MEC for CM in walnuts was evaluated over several seasons in California funded by the California Walnut IPM Expansion Project (**Table 1**) [36]. Unfortunately, studies were often conducted in groves with low pest pressure and results were inconclusive. Studies evaluated rates of 10, 20, and 30 g a.i., added stickers (0.06%), and made five applications per season every four weeks using 940 L ha^{-1}. Studies compared 3 M-CM and CheckMate® CM-F and they found that both could reduce male catches in sex pheromone-baited traps, but nut injury was often not reduced [52]. MD programs in walnuts were also much more expensive than current conventional programs which required only a few sprays per season [51]. Finally, the occurrence of the navel orangeworm, *Amyelois transitella* (Walker) (Lepidoptera: Pyralidae), as a secondary direct pest also made the use of a selective tactic for a single pest too expensive.

Several studies reported poor results with MEC formulations for CM in apples using air blast sprayers and standard water volumes. Comparative studies found moderate reductions in male catches with sprayable MD, but neither the 3 M-CM nor CheckMate® CM-F reduced CM injury [53]. Substantial variations in the efficacy of MEC trials with CM were reported suggesting they were not reliable [54]. Similarly, CheckMate® CM-F was never as effective as hand-applied dispensers in shutting down traps with 25–40% less disruption in 1st flight and 75% less disruptive later in the season [55].

Despite all of these negative CM trials, the recognition of the superior efficacy of using a low-volume application of CheckMate® CM-F was never acknowledged or tested by these researchers [56, 57]. The most extensive studies of low volume CheckMate® CM-F applications were conducted from 2004 to 2005 in 20 apple and pear grower's orchards in WA and OR (**Table 1**) [37]. These studies consistently demonstrated two results: (*i*) higher moth catches could occur in low volume CheckMate® CM-F-treated orchards than similar blocks treated with hand-applied dispensers; (*ii*) levels of fruit injury were never significantly different between hand-applied dispensers and sprayable MEC (**Table 1**). The combined use of CheckMate® CM-F with CIDETRAK® DA MEC™ was effective at both low and high spray volumes in Türkiye (**Table 1**) [28].

Similar but more limited number of studies have been conducted with low volume applications of CheckMate® OFM-F in California and Türkiye in stone fruits [50]. The California studies compared three rates of five applications beginning in March against dispensers and untreated plots and the 2006 studies used one rate and started in May with four applications. The study in Türkiye compared a low volume and a high-volume applications (3 sprays beginning in June) using an air blast sprayer but with some nozzles turned off in the low volume application. The data consistently found that while moth counts of OFM were higher in plots treated with the MEC formulations than dispensers, the fruit injury was either lower or not significantly different from dispenser treatments [50].

Measuring the impact of MD on the female moth mating status is a more direct measurement of the efficacy of any product [58]. Unfortunately, there is not a lot of data on the effect of MEC formulations on females mating success. Only three CM studies each using a different sampling method have been published. Tethered females in 3 M-CM-treated plots had 80% reduction in mating [17], which is comparatively low compared with various dispenser studies [58]. VF-baited traps were used to evaluate the comparative efficacy of low volume sprays and retained 75% male catch reductions after three weeks compared to no effect with standard spray volumes (**Table 1**) [32]. These two passive monitoring methods use fixed-positions for laboratory-reared virgin females and can overestimate the disruption of mating by wild and mobile females [58]. The third study reported the proportion of wild CM females trapped that were mated was somewhat lower in treated (CheckMate® CM-F + CIDETRAK® DA MEC™) than untreated plots; and was lower with the low volume compared with the standard spray volume (**Table 1**) [38].

2.2 Low volume applications: Most effective use of MEC

It is interesting that the most effective use of the CheckMate® CM-F (25 g a.i. in <20 L ha^{-1}) was developed by accident as a means for the researcher to reduce the time needed to set up distant studies and still visit aging parents before returning to the office [35]. However, these positive results, which seemed like an anomaly at the time when compared with other more negative reports of MEC efficacy, were clearly demonstrated with a comparative spray volume test in 2000 (**Table 1**) [32]. The low volume application was shown to both significantly increase the deposition of capsules in the canopy and reduce levels of fruit injury. In fact, spraying the CheckMate® CM-F with a standard spray volume was no more effective than spraying the material on the ground [32]. Additional supporting trials were conducted in 2005 and 2006 and found that the low volume sprays with CheckMate® CM-F were equal to hand-applied dispensers in both apple and pear blocks [37, 59]. A similar study confirmed the improved efficacy of low volume applications of OFM-F in stone fruits in California and Türkiye [50]. The low volume studies with CheckMate® CM-F in Türkiye adjusted spray volume by turning off some nozzles on the air blast sprayer [50, 60]. Male counts were lower, and capsule density was higher with the low versus higher volumes. Similar results were reported in Türkiye with CheckMate® CM-F in apple (**Table 1**) [28].

2.3 Mechanism of low volume MEC sprays

The uniform broadcast spray application of MEC formulations was thought to be effective due to a noncompetitive camouflaging mechanism of disruption [7]. The increase deposition of capsules with low-volume applications thus likely enhances this effect. This initial mechanism for mating disruption was supported by field data [27]. Moth catches in low-volume sprayed blocks in traps with synthetic lures were initially low but significantly increased in weeks 3 and 4. Studies suggested that charged leaves could act as attractive point sources for an extended period [27, 61, 62]. For example, field-aged leaves with ca. 100 capsules were attractive to males flown in a flight tunnel for 5 weeks [61]. Similarly, traps baited with MEC-treated artificial leaves were attractive for at least 5 weeks in the field [27]. These data suggest that MEC applications initially disrupt males through camouflage (initial burst of pheromone) for a relatively short period and then this is followed by an extended period of competitive attraction caused by thousands of point sources (leaves with clumps of capsules). A few other studies have considered the role of individual leaves treated with MEC formulations on moth behavior. Grapevine, *Vitis* spp., leaves served as passive dispensers for *Lobesia botrana* (Denis and Schiffmüller) (Lepidoptera: Tortricidae), for two weeks [62]. Some of the activity of the 3 M formulations for the tortricid leafrollers *Choristoneura rosaceana* (Harris) and *Pandemis limitata* (Robinson) were hypothesized to be greater if moths resting on treated leaves [44].

2.4 New MEC technology

The new Trécé Inc. MEC formulation to date has mostly been evaluated with pear ester. The microcapsules are much smaller than the CheckMate® CM-F microbeads (mean of 10 versus 100 μm) and significantly impacted CM larval wandering behavior on leaves for 14 d [26]. The impact of its much smaller capsule size on effective deposition and retention has not been studied. CIDETRAK® DA MEC™ is widely used by walnut and apple growers to enhance insecticide-based control [25, 63]. CIDETRAK® DA MEC™ significantly reduced levels of injury when used in combination with codlemone or codlemone/pear ester dispensers [64].

Comparable evaluations of the CIDETRAK® CM MEC™ product have not been as well reported [65, 66]. Similar to the earlier MEC formulations for CM, CIDETRAK® CM MEC™ does not disrupt male catches as well as other hand-applied dispensers [66]. The use of low volume applications with this formulation has not been reported. Herein, we report comparative laboratory and field data on the comparative efficacy of a pre-registration Trécé Inc. formulation (TRE1505) with CheckMate® CM-F from 2015 to 2018 and field trials with CIDETRAK® CM MEC™ from 2020 to 2024 tested against several hand-applied dispensers.

3. Materials and methods

3.1 Field sprays / flight tunnel assays

Studies were conducted from 2015 to 2018 to evaluate the experimental CIDETRAK® TRE1505 prior to its registration as CIDETRAK® CM MEC™ in 2019 to determine the optimal rate and factors impacting its longevity. In particular, the new formulation was compared with the only remaining commercial product, CheckMate® CM-F. Trials 1 and 2 evaluated the attraction of laboratory-reared male CM to apple leaves sprayed with each formulation aged over eight weeks, and the effect of overhead watering with the use of a latex sticker (0.06%, Bond Max, Loveland Industries, Greely, CO) over four weeks. Potted 10-year-old 'Delicious' trees (ca. 2.2 m height) were sprayed with each formulation (N = 5) using a backpack power sprayer at a standard rate of 10 g a.i. in 12 L water ha^{-1} until drip in both trials. Leaves were collected each week for eight weeks and used in a flight tunnel protocol previously described [27]. Trial 2 placed the sprayed potted trees under an overhead watering system, delivering 120 L per h every day for 4 hours (1400 – 1800 h) for four weeks. Trial 3 compared the residual attractiveness of field-collected apple leaves treated with either CheckMate® CM-F or TRE1505. Either the upper or lower apple leaf surfaces were sprayed with a rain simulator (Model SB6-059, DeVries Manuf., Hollandale, MN) as described previously [27]. Data for the proportion of male moths orienting within 10 cm of the leaf and touching the leaf were recorded in both trials. The specific flight tunnel assay with acclimated laboratory-reared male CM was described previously [27]. Male response to untreated leaves were included in each of the three trials and no males crossed the mid-line of the flight tunnel.

3.2 Field trials of TRE1505, 2016–2018

Five field studies compared the two formulations in conventional apple orchards situated near Wapato, WA (46.3-46.6° N, 120.2-120.4° W). Treatments included air blast (935 L ha^{-1}) and low volume (25 L ha^{-1}) applications with and without a latex sticker, 0.06%. The low volume sprayer was the apparatus used by cherry growers to apply the bait GF-120 for the western cherry fruit fly, *Rhagoletis indifferens* (Curran) [67]. Trial 7 included plots treated with CIDETRAK® CMDA COMBO™ PP at 1000 dispensers ha^{-1} loaded with 85 and 50 mg of codlemone and pear ester. Replicated (N = 4 to 8) 1-ha plots were randomized across four contiguous 8-ha blocks of apple in each trial. Similar untreated blocks were included in three trials. All blocks were sprayed with similar conventional spray programs each year. Each replicate was monitored with delta traps baited with 0.1 and 1.0-mg codlemone lures. Lures and trap liners were replaced each week. Studies were conducted for ca. four weeks from early July or August to late August or early September during the 2nd and 3rd flight of CM. All blocks were treated with conventional spray programs that were likely less than required and levels of CM fruit injury were similar across blocks each year and ranged from ca. 1.0 to 3.0%. Accumulated field precipitation was recorded during each trial.

3.3 Field trials of CIDETRAK® CM MEC™, 2020-2024

Seasonal field studies were conducted from 2020 to 2024 with CIDETRAK® CM MEC™ in comparison with either the CIDETRAK® CMDA COMBO™ PP applied at 700 dispensers ha^{-1} in 2020–2021 and 2023–2024 or the CIDETRAK® CMDA COMBO™ MESO™-A loaded with 850 and 500 mg of codlemone and pear ester and applied at 80 dispensers ha^{-1} in 2022. Replicated plots each year were 2 to 4 ha. The study in 2021 randomly split six orchards into two 2.0-ha plots. Studies in 2023–2024 compared the use of CIDETRAK® CM MEC™ alone and in combination with the CIDETRAK® CMDA COMBO™ PP dispensers. Four or five applications (spaced 4–5 weeks apart) of sprayable MEC at 25 g a.i. ha^{-1} with an air blast sprayer delivering 935 L ha^{-1} were made starting in early May each year, except in 2023 when only one application was made on 11 July and the study lasted four weeks. Dispensers were applied to the upper third of the canopy in mid-April. Monitoring traps baited with either 1.0 mg codlemone or a 4-component non-pheromone lure, PHEROCON® MEGALURE CM DUAL 4 K (Trécé Inc.) were placed in the last week of April and checked every two weeks. *The Coefficient of Male MD* [65] was calculated each year as: (male catch in traps with the 1.0 mg pheromone lure divided by male catch in traps with the 4 K lure) for each treatment, except in 2023. Female CM collected from the various orchards were dissected to determine their mating status. Females from untreated plots in 2021 were from nearby insecticide-only treated orchards.

Studies in 2023 and 2024 included weekly releases of unsexed sterilized CM obtained from the Okanagan-Kootenay Sterile Insect Release Program's mass rearing facility (Osoyoos, BC, Canada). A single 150-m transect was established in each orchard and five traps with the PHEROCON® MEGALURE CM DUAL 4 K lure were spaced at 30 m. Approximately 400 unsexed sterile CM were released down each row each week by lightly tapping a chilled moth container while driving down the row in an ATV. Liners were collected seven days later prior to the next moth release. Moths were sexed and all sterile and wild females were dissected to determine their mating status using a stereomicroscope. All female CM caught in traps with dual sex lures in each treatment replicate were also dissected. All orchards were also treated with a conventional spray program and levels of fruit injury were less than 0.1% and are not presented.

3.4 Statistical analyses

All count and proportional data were transformed with sqrt(x) and arcsin(sqrt(x)) prior to an analysis of the normality of the data (Shapiro-Wilk). Normalized untransformed or transformed data were evaluated with a one- or two-way analysis of variance (F-test statistic). Tukey's method was used to separate means in significant ANOVAs, $P < 0.05$. The Kruskal-Wallis nonparametric test (W-test statistic) was used to analyze nonnormal data sets. This was followed by an ANOVA applied to ranks. Fisher's exact test was used to compare the proportion of male moths approaching and contacting sprayed leaves in a flight tunnel [68].

4. Results

4.1 Field sprays / flight tunnel assays

TRE1505 outperformed the CheckMate® CM-F formulation in Trial 1 in terms of the proportion of males flying within 10 cm of the source or directly touching

Trial # dates precipitation during trial	**Treatment**	**# males flown per date**	**Proportion males in 29-d bioassay**		**Last post-spray bioassay with >50% male response**	
			Flying close to source	**Touching source**	**Flying close to source**	**Touching source**
1 19 May – 22 July 2015 8.1 mm	CheckMate® CM-F	15	0.40 B	0.40 B	day 22	day 22
	CIDETRAK® TRE1505	15	1.00 A	0.80 A	day 57	day 57
2 11 July – 9 August 2016 2.5 mm	CIDETRAK® TRE1505 + sticker	12	0.83 A	0.67 A	> day 29	> day 29
	CIDETRAK® TRE1505 + sticker Under water	12	0.75 A	0.58 A	> day 29	> day 29

Column means in each trial followed by a different letter were significantly different, Fisher's Exact test, $P < 0.05$.

Table 2.
Comparison of male CM flight response (flying <10 cm from leaf and touching leaf) in weekly assays with field aged sprayed apple leaves treated with or without a pine resin sticker and with or without overhead daily watering (1400–1800 h), 2015–2016.

the sprayed leaves in the 57-d (**Table 2**). Males approached 100% of leaves treated with TRE1505 at day 29 greater than 50% of males were still touching sprayed leaves at day 57. The use of a pine resin sticker in Trial 2 did not extend male responses to treated leaves subjected to four hours of overhead watering through 29 days.

Significant differences in the responses of male CM were flown in a flight tunnel to apple leaves sprayed with each formulation to either the top or bottom surface of apple leaves with or without the latex sticker (**Table 3**). A greater proportion of males responded to (flying close and touching) leaves treated with TRE1505 than with CheckMate® CM-F. The addition of the sticker increased the response of males to either top or bottom treated leaves under water to levels similar to treated leaves not watered with either formulation.

4.2 Field trials of TRE1505, 2016-2018

Significant differences in male CM catch in traps baited with 0.1 or 1.0 mg codlemone occurred in four of the five four-week field trials (**Table 4**). Moth counts were lower with both lures with TRE1505 with low volume versus air blast applications in Trial 4. Moth counts with the air blast applications of CheckMate® CM-F were intermediate. Trial 5 found that low volume applications of both formulations significantly reduced male counts with either lure compared with an untreated control. Trial 6 found no difference in moth catches with either lure in plots treated with TRE1505 used with or without a latex sticker. Significant differences in male catches between low volume sprayable formulations and hand-applied dispenser occurred in Trial 7. Mean catch was not different with the 0.1 mg lure between TRE1505 and dispensers but was lower in dispenser-treated replicates with the 1.0-mg lure. Male catches were not different between the two MEC formulations. Finally, Trial 8 found

Trial #	MEC type	Treatment			Mean ± SE proportion of males	
		Watering	Leaf surface watered	Sticker added	Flying close to source	Touching source
3	CheckMate® CM-F	Yes	Top	no	0.50 ± 0.07 bB	0.15 ± 0.08 bB
		Yes	Bottom	no	0.51 ± 0.10 bB	0.27 ± 0.09 bB
		Yes	Top	yes	0.90 ± 0.06 aB	0.52 ± 0.07 aB
		Yes	Bottom	yes	0.92 ± 0.08 aB	0.75 ± 0.08 aB
		No	—	no	0.99 ± 0.01 aB	0.81 ± 0.05 aB
	CIDETRAK® TRE1505	Yes	Top	no	0.67 ± 0.07 bA	0.27 ± 0.08 bA
		Yes	Bottom	no	0.78 ± 0.06 bA	0.56 ± 0.07 bA
		Yes	Top	yes	0.95 ± 0.05 aA	0.81 ± 0.07 aA
		Yes	Bottom	yes	1.00 ± 0.00 aA	0.78 ± 0.11 aA
		No	—	no	1.0 ± 0.00 aA	0.86 ± 0.04 aA
Statistics				MEC	$F_{1,63} = 8.60$ $P = 0.0047$	$F_{1,63} = 7.41$ $P = 0.0084$
				Treatment	$F_{4,63} = 26.51$ $P < 0.0001$	$F_{4,63} = 24.36$ $P < 0.0001$
				MEC x Treatment	$F_{4,63} = 1.10$ $P = 0.3643$	$F_{4,63} = 0.86$ $P = 0.4932$

Column means followed a different uppercase or lowercase letter were significantly different (P < 0.05) for MEC and treatment (water, surface, sticker), respectively.

Table 3.
Flight tunnel evaluations of male CM responding to MEC-treated apple leaves exposed to simulated rainfall, 2015.

Trial #	Dates replicates / precipitation	Treatment, sprayer	Mean ± SEM male catch per trap	
			0.1 mg lure	1.0 mg lure
4	13 July – 16 Aug. 2016 N = 6 / 2.5 mm	CM-F, air blast	3.8 ± 1.4 AB	13.5 ± 2.8 AB
		TRE1505, low volume	1.8 ± 1.2 B	10.7 ± 1.8 B
		TRE1505, air blast	6.7 ± 1.3 A	26.0 ± 5.9 A
		Statistics	$F_{2,15} = 3.51$ $P = 0.0563$	$F_{2,15} = 4.60$ $P = 0.0276$
5	25 July – 23 Aug. 2016 N = 4 / 0.8 mm	CM-F, low volume	0.8 ± 0.5 B	9.5 ± 4.2 B
		TRE1505, low volume	0.3 ± 0.3 B	14.5 ± 4.4 B
		Untreated	6.0 ± 2.8 A	58.0 ± 5.9 A
		Statistics	$F_{2,9} = 9.11$ $P = 0.0069$	$F_{2,9} = 17.91$ $P = 0.0007$
6	11 Aug. – 9 Sept. 2016 N = 4 / 1.3 mm	TRE1505, low volume	5.0 ± 1.1 A	47.0 ± 18.5 A
		TRE1505 + sticker, low volume	4.0 ± 1.2 A	32.8 ± 12.6 A
		Statistics	$F_{1,6} = 0.40$ $P = 0.5511$	$F_{1,6} = 0.25$ $P = 0.6328$

Trial #	Dates replicates / precipitation	Treatment, sprayer	Mean ± SEM male catch per trap	
			0.1 mg lure	1.0 mg lure
7	7 July – 6 Sept. 2017 N = 6 / 0.3 mm	CM-F, sticker, low volume	12.0 ± 1.4 AB	25.3 ± 4.1 B
		TRE1505, sticker low volume	4.8 ± 0.7 B	21.3 ± 3.5 B
		Cidetrak CMDA PP dispenser	3.5 ± 0.2 B	5.5 ± 0.6 C
		Untreated	25.2 ± 3.7 A	75.2 ± 9.7 A
		Statistics	$W = 19.56$ $P = 0.0002$	$F_{3, 20} = 36.88$ $P < 0.0001$
8	7 Aug. – 3 Sept. 2018 N = 8 0.0 mm	TRE1505, sticker air blast	5.9 ± 1.0 B	28.6 ± 3.3 B
		TRE1505, sticker low volume	2.5 ± 0.5 B	10.9 ± 2.0 C
		Untreated	40.3 ± 6.6 A	89.1 ± 9.0 A
		Statistics	$W = 17.76$ $P = 0.0001$	$F_{2, 21} = 76.47$ $P < 0.0001$

The F-statistic was calculated in ANOVAs with normalized data (Shapiro-Wilks test) and column means in each trial followed by a different letter were significantly different, P < 0.05, Tukey HSD test. Kruskal-Wallis (W-statistic) was used if the data could not be normalized. A parametric ANOVA of ranks was then calculated, and means were separated at P < 0.05.

Table 4.
Comparison of male CM catches in delta traps baited with either 0.1 or 1.0 mg septa sex pheromone lures in small-plot (1 ha) replicated four-week MD field trials with CheckMate® CM-F or the CIDETRAK® TRE1505 formulations applied at 25 g a.i. ha^{-1} with low volume 24 L or air blast 935 L ha^{-1} sprays, 2016–2018.

that the low volume application of TRE1505 was significantly lower than the air blast application with the 1.0 mg but not the 0.1 mg lure. Both sprayable treatments significantly reduced moth catches compared with the untreated plots. Accumulated precipitation during each trial was low, 0.0 to 2.5 mm.

4.3 Field trials of CIDETRAK® CM MEC™, 2020-2024

Trials 9–13 conducted over five years completed a number of different comparisons with CIDETRAK® CM MEC™ was compared with dispensers and in

Trial # Dates Year Precipitation	Treatment [a]	No. of replicates	Mean ± SE value of *Coefficient of Male MD*	Mean ± SE proportion of females not mated [# females dissected]
Trial 9 24 April - 22 Aug. 2020 29 mm	Untreated	6	1.56 ± 0.30 A	0.23 ± 0.03 B [267]
	CM-MEC	6	0.66 ± 0.16 AB	0.43 ± 0.05 A [194]
	PP dispensers	6	0.18 ± 0.04 B	0.40 ± 0.08 AB [143]
	Statistics		$W = 13.15$ $P = 0.0014$	$F_{2, 15} = 4.00$ $P = 0.0405$
Trial 10 28 April - 8 Sept. 2021 4 mm	UTC	6	—	0.27 ± 0.11 B [131]
	CM-MEC	6	1.09 ± 0.27 A	0.68 ± 0.09 A [70]
	PP dispensers	6	0.33 ± 0.19 B	0.80 ± 0.06 A [121]
	Statistics		$W = 7.17$ $P = 0.0074$	$F_{2, 15} = 7.48$ $P = 0.0056$

Trial # Dates Year Precipitation	Treatment [a]	No. of replicates	Mean ± SE value of *Coefficient of Male MD*	Mean ± SE proportion of females not mated [# females dissected]
Trial 11 23 April - 3 Sept. 2022 28 mm	UTC	4	0.72 ± 0.14 A	0.28 ± 0.05 B [55]
	CM-MEC	5	0.86 ± 0.41 A	0.70 ± 0.10 A [82]
	Meso dispensers	8	0.35 ± 0.09 B	0.45 ± 0.08 AB [299]
	Statistics		$W = 15.35$ $P = 0.0015$	$F_{2, 14} = 4.06$ $P = 0.0406$
Trial 12 11 July - 8 Aug. 2023 20 mm	UTC	4	—	Wild: 0.28 ± 0.12 A [35] Sterile: 0.31 ± 0.03 C [340]
	CM-MEC	4	—	Wild: 0.33 ± 0.08 A [63] Sterile: 0.31 ± 0.08 BC [270]
	Dispensers	4	—	Wild: 0.23 ± 0.11 A [27] Sterile: 0.59 ± 0.04 AB [19]
	CM-MEC + PP dispensers	4	—	Wild: 0.27 ± 0.05 A [184] Sterile: 0.68 ± 0.12 A [158]
	Statistics		-	Wild: $W = 2.27$, $P = 0.5188$ Sterile: $F_{3, 45} = 7.69$, $P = 0.0004$
Trial 13 13 May - 7 Sept. 2024 8 mm	UTC	4	Wild: 2.00 ± 0.28 A Sterile: 1.71 ± 0.21 A	Wild: 0.17 ± 0.08 B [105] Sterile: 0.35 ± 0.01 B [955]
	CM-MEC	4	Wild: 0.85 ± 0.18 AB Sterile: 0.72 ± 0.20 AB	Wild: 0.40 ± 0.03 AB [446] Sterile: 0.60 ± 0.03 A [348]
	PP dispensers	4	Wild: 0.15 ± 0.06 B Sterile: 0.35 ± 0.11 B	Wild: 0.32 ± 0.05 B [120] Sterile: 0.33 ± 0.08 B [165]
	CM-MEC + PP dispensers	2	Wild: 0.10 ± 0.03 B Sterile: 0.26 ± 0.08 B	Wild: 0.77 ± 0.03 A [130] Sterile: 0.93 ± 0.07 A [91]
	Statistics		Wild: $F_{3, 10} = 4.31$, $P = 0.0154$ Sterile: $F_{3, 10} = 7.69$, $P = 0.0004$	Wild: $F_{3, 10} = 8.53$, $P = 0.0041$ Sterile: $F_{3, 10} = 11.93$, $P = 0.0017$

[a]Treatments included CIDETRAK® CM MEC™ and the CIDETRAK® CMDA COMBO™ PP and CIDETRAK® CMDA COMBO™ MESO™-A dispensers.

Table 5.
Field studies in replicated conventional apple orchards used to calculate the Coefficient of Male MD *defined as the ratio of male catch in traps baited with a PHEROCON® CM 1.0 mg lure divided by male catch in traps with the non-pheromone lure, PHEROCON® MEGALURE CM DUAL 4 K and the proportion of trapped female CM that were not mated in apple orchards either untreated, sprayed with CIDETRAK® CM MEC™ or treated with dispensers loaded with codlemone / pear ester from 2020 to 2024.*

combination with dispensers (**Table 5**). Calculations of the *Male Coefficient of MD* confirmed that the CIDETRAK® CM MEC™ does not disrupt pheromone-baited traps better than untreated plots and significantly lower than dispensers. However, the proportions of unmated female CM were similar between the sprayable and dispensers in each trial, except in 2024 where CIDETRAK® CM MEC™ outperformed dispensers in disrupting the sterile females (**Table 5**). The combination of spraying CIDETRAK® CM MEC™ in dispenser-treated orchards was more effective than the sprayable alone in 2023 and dispensers alone in 2024.

5. Discussion

The CIDETRAK® CM MEC™ formulation was found to be more effective than CheckMate® CM-F in semi-field experiments due to extended activity and greater retention despite overhead watering and was significantly improved with the use of a sticker. The two MEC formulations differ 10-fold in their mean capsule size, and we hypothesized that the smaller capsule size with CIDETRAK® CM MEC™ is a key factor improving its performance. To date, useful studies characterizing microcapsule densities and distribution using UV dyes have not been done with the CIDETRAK® CM MEC™ formulation. In addition, the characterization of leaves with variable microcapsule densities acting as attractive point sources with this formulation has not been conducted. The improvement in Trials 4 and 8 with low volume applications of TRE1505 suggests that this application method works similarly to our earlier studies with CheckMate® CM-F. At present apple growers do not apply any pesticides with ATVs except pollen or herbicide spot treatments, so it is uncertain if growers would be willing to apply low volume MEC sprays. The lower treatment cost using the ATV likely allows the MEC sprays to be applied independent from other mixed tank applications and would improve timings.

Apparently, MEC formulations do not disrupt male CM catch in pheromone-baited traps as well as the use of dispensers. In fact, male catch may be increased due to the MEC spray [66]. This failure to reduce male catch in sex pheromone-baited traps has been interpreted as a failure to disrupt mating [58]. However, female dissections tell a different and more positive story concerning the efficacy of MEC-based programs. The greater levels of disruption found with use of MEC in combination with dispensers can provide a new approach to manage CM: supplemental MEC sprays in orchards with a basic MD dispenser framework. One or two MEC sprays per generation could be used depending on pest density.

6. Conclusion

Research over the past 10 years with MEC sex pheromone formulations for CM has made several new discoveries demonstrating a significant improvement in their effectiveness. Factors that can increase the deposition and retention of microcapsules are key. Retention of the 1.0 ug microcapsules despite precipitation, irrigation or overhead watering is much better than with the larger microbeads. Their deposition using air blast applications was relatively ineffective. Low volume applications with less water and much lower wind speed can significantly improve the performance of the smaller microcapsules.

Several options exist for the effective use of the new MEC formulation. Season-long monthly applications is one approach. Reducing the density of dispensers or aerosols and applying the MEC only at generational flight peaks during the season is another. Adjusting the number of seasonal MEC sprays based on moth densities estimated with the *Coefficient of Male MD* can be used to optimize programs.

Author details

Alan Lee Knight[1], Esteban Basoalto[2]* and Michele Preti[3]

1 Instar Biologicals, Yakima, Washington, USA

2 Facultad de Ciencias Agrarias y Alimentarias, Universidad Austral de Chile, Valdivia, Chile

3 Integrated Pest Management Consultant and Researcher, Faenza, Italy

*Address all correspondence to: esteban.basoalto@uach.cl

References

[1] Shorey HH, Kaae RS, Gaston LK, McLaughlin. Sex pheromones of lepidoptera. XXX. Disruption of sex pheromone communication in *Trichoplusia ni* as a possible means of mating control. Environmental Entomology. 1972;**1**(5):641-645. DOI: 10.1093/ee/1.5.641

[2] Doane CC, Brooks TW. Research and development of pheromones for insect control with emphasis on the pink bollworm. In: Mitchell ER, editor. Management of Insect Pests with Semiochemicals. Boston, MA: Springer; 1981. DOI: 10.1007/978-1-4613-3216-9_24

[3] Cardé RT, Baker TC, Castrovillo PJ. Disruption of sexual communication in *Laspeyresia pomonella* (codling moth), *Grapholita molesta* (oriental fruit moth) and *G. Prunivora* (lesser apple worm) with hollow fiber attractant sources. Entomologia Experimentalis et Applicata. 1977;**22**:280-288. DOI: 10.1111/j.1570-7458.1977.tb02718.x

[4] Moffitt HR, Westigard PH. Suppression of the codling moth (Lepidoptera; Tortricidae) populations on pears in southern Oregon through mating disruption with sex pheromone. Journal of Economic Entomology. 1984;**77**(6):1513-1519. DOI: 10.1093/jee/77.6.1513

[5] Stelinski LL, Miller JR, Ledebuhr R, Siegert P, Gut LJ. Season-long mating disruption of *Grapholita molesta* (Lepidoptera: Tortricidae) by one machine application of pheromone in wax drops (SPLAT-OFM). Journal of Pest Science. 2007;**80**:109-117. DOI: 10.1007/s10340-007-0162-0

[6] Gut LJ, Adams C, Miller J, McGhee P, Thomson D. Biological control in integrated management of deciduous fruit insect pests: The use of semiochemicals. In: Xu X, Fountain M, editors. Integrated Management of Diseases and Insect Pests of Tree Fruit. Philadelphia: Burleigh Dodds Sci. Publishing; 2019. pp. 571-662. eBook: ISBN9780429266690

[7] Hall DR, Nesbitt BF, Marrs GJ, Green AJ, Campion DG, Critchley BR. Development of microencapsulated pheromone formulations. In: Leonhardt BA, Beroza M, editors. Insect Pheromone Technology: Chemistry and Applications. Vol. 190. Washington, DC: ACS Publications; 1982. pp. 131-143. DOI: 10.1021/bk-1982-0190.ch006

[8] Gu XL, Zhu X, Kong XZ, Tan Y. Comparisons of simple and complex coacervations for preparation of sprayable insect sex pheromone microcapsules and release control of the encapsulated pheromone molecule. Journal of Microencapsulation. 2010;**27**(4):355-364. DOI: 10.3109/02652040903221532

[9] Yusop FH, Manaf SFA, Hamzah F. Preservation of bioactive compounds via microencapsulation. Chemical Engineering Research Bulletin. 2017;**19**:50. DOI: 10.3329/cerb.v19i0.33796

[10] Hellmann C, Greiner A, Vilcinskas A. Design of polymer carriers for optimized pheromone release in sustainable insect control strategies. Advancement of Science. 2024;**11**(9):e2304098. DOI: 10.1002/advs.202304098

[11] Sun C, Wang A, Shen Y, Li X, Zhan S, Wang C, et al. Innovative matrices to release pheromones for integrated pest management. Entomologia Generalis.

2024;**44**:1135-1144. DOI: 10.1127/entomologia/2024/2772

[12] Schwalbe CP, Camerson EA, Hall DJ, Richerson JV, Beroza M, Stevens LJ. Field tests of microencapsulated disparlure for suppression of mating among wild and laboratory-reared gypsy moths. Environmental Entomology. 1974;**3**(4):589-592. DOI: 10.1093/ee/3.4.589

[13] Grenzebach J, Larsen T. Laboratory evaluations of Consep's sprayable pheromone formulations. In: Proceedings of the 74th Annual Western Orchard Pest and Disease Management Conference; 5-7 January 2000. Portland, OR: WSU; 2000. p. 89

[14] Muir D, Peatman C, Sawka M. 3m TM sprayable pheromones. In: Proceedings of the 73rd Annual Western Orchard Pest and Disease Management Conference; 6-8 January 1999. Portland, OR: WSU; 1999. p. 115

[15] Athanassiou CG, Kavallieratos NG, Benelli G, Losic D, Usha Rani P, Desneux N. Nanoparticles for pest control: Current status and future perspectives. Journal of Pest Science. 2018;**91**:1-15. DOI: 10.1007/s10340-017-0898-0

[16] Polavarapu S, Lonergan G, Peng H, Neilsen K. Potential for mating disruption of *Sparganothis sulfureana* (Lepidoptera: Tortricidae) in cranberries. Journal of Economic Entomology. 2001;**94**(3):658-665. DOI: 10.1603/0022-0493-94.3.658

[17] Charmillot P-J, Pasquier D. Essai preliminaire de lutte par confusion contre la cochylis *Eupoecilia ambiguella* et le carpocapse *Cydia pomonella* au moyen des microcapsules 3M. IOBC WPRS Bulletin. 2001;**24**(2):63-64

[18] Taschenberg EF, Roelofs WL. Pheromone communication disruption of the grape berry moth with microencapsulated and hollow fiber systems. Environmental Entomology. 1976;**5**:688-691. DOI: 10.1093/ee/5.4.688

[19] Trimble RM, Vickers PM, Nielsen KE, Barinshteyn G. Sprayable pheromone for controlling the north American grape berry moth by mating disruption. Agricultural and Forest Entomology. 2003;**5**(3):263-268. DOI: 10.1046/j.1461-9563.2003.00188.x

[20] Cardé RT, Trammel K, Roelofs WL. Disruption of sex attraction of the redbanded leafroller (*Argyrotaenia velutinana*) with microencapsulated pheromone components. Environmental Entomology. 1975;**4**:448-450. DOI: 10.1093/ee/4.3.448

[21] Stelinski LL, McGhee P, Hua SM, Il'ichev AL, Gut LJ. Sprayable microencapsulated sex pheromone formulations for mating disruption of 4 tortricid species: Effect of application height, rate, frequency, and sticker adjuvant. Journal of Economic Entomology. 2007;**100**(4):1360-1369. DOI: 10.1603/0022-0493

[22] Agrian label [Internet]. 2025. Available from: https://www.agrian.com/labelcenter/resources.com [Accessed: April 25, 2025]

[23] Knight AL, Larsen TE, Ketner KC. Rainfastness of a microencapsulated sex pheromone formulation for codling moth (Lepidoptera: Tortricidae). Journal of Economic Entomology. 2004;**97**(6):1987-1992. DOI: 10.1603/0022-0493-97.6.1987

[24] Light DM, Knight AL. Kairomone-augmented mating disruption control for codling moth in Californian walnuts and apples. IOBC-WPRS Bulletin. 2005;**28**:341-344

[25] Knight AL, Light DM. Adding microencapsulated pear ester to

insecticides for control of *Cydia pomonella* (Lepidoptera: Tortricidae) in apple. Pest Management Science. 2012;**69**:66-74. DOI: 10.1002/ps.3363

[26] Light DM, Beck JJ. Characterization of microencapsulated pear ester, (2*E*,4*Z*)-Ethyl-2,4-decadienoate, a kairomonal spray adjuvant against neonate codling moth larvae. Journal of Agricultural and Food Chemistry. 2010;**58**:7838-7845. DOI: 10.1021/jf101167p

[27] Knight AL, Larsen TE. Creating point sources for codling moth (Lepidoptera: Tortricidae) with low-volume sprays of a microencapsulated sex pheromone formulation. Environmental Entomology. 2008;**37**:1136-1144. DOI: 10.1093/ee/37.5.1136

[28] Kovanci OB, Kumral NA, Larsen TE. High versus ultra-low volume spraying of a microencapsulated pheromone formulation for codling moth control in two apple cultivars. International Journal of Pest Management. 2010;**56**(1):1-7. DOI: 10.1080/09670870902957279

[29] Casagrande E. Use of sprayable pheromone formulations in Europe. IOBC-WPRS Bulletin. 2010;**54**(2):411-413. ISBN 978-92-9067-289-0

[30] Waldstein DE, Gut LJ. Comparison of microcapsule density with various apple tissues and formulations of oriental fruit moth (Lepidoptera: Tortricidae) sprayable pheromone. Journal of Economic Entomology. 2003;**96**:58-63. DOI: 10.1093/jee/96.1.58

[31] Waldstein DE, Kering MK. Studies on oriental fruit moth (Lepidoptera: Tortricidae) pheromone microcapsules using various fruit species, cultivars, and application methods. Journal of Agricultural and Urban Entomology. 2006;**23**(1):7-16

[32] Knight AL, Larsen TE. Improved deposition and performance of a microencapsulated sex pheromone formulation for codling moth (Lepidoptera: Tortricidae) with a low volume application. Journal of the Entomological Society of British Columbia. 2004;**101**:79-86

[33] Racsko J, Schrader LE. Sunburn of apple fruit: Historical background, recent advances and future perspectives. Critical Reviews in Plant Sciences. 2012;**31**(6):455-504. DOI: 10.1080/07352689.2012.696453

[34] Ketner K, Eng J, Holmes E, Larsen T. Effect of rainfall on encapsulated codling moth pheromone. In: Proceedings of the 77th Annual Western Orchard Pest and Disease Management Conference; 15-17 January 2003. Portland, OR: WSU; 2003. p. 10

[35] Larsen TL. Field studies with encapsulated sprayable codling moth sex pheromone. In: Proceedings of the 76th Annual Western Orchard Pest and Disease Management Conference; 9-11 January 2002. Portland, OR: WSU; 2002. p. 22

[36] Grant JA, Pickel CP, Van Steenwyk RA, Welter SC. Management of codling moth using sprayable microencapsulated pheromones with selective insecticides. California Walnut Board Report [Internet]. 2004. Available from: https://my.ucanr.edu/repository/fileAccessPublic.cfm?fn=2004_195_ocr.pdf-154369.pdf [Accessed: April 25, 2025]

[37] Knight AL, Larson TL, Ketner K, Hilton R, Hawkins L. Field evaluations of concentrated spray applications of a microencapsulated sex pheromone formulation for codling moth (Lepidoptera: Tortricidae). Environmental Entomology.

2008;**37**(4):980-989. DOI: 10.1093/ee/37.4.980

[38] Kovanci OB. Co-application of microencapsulated pear ester and codlemone for mating disruption of *Cydia pomonella*. Journal of Pest Science. 2015;**88**:311-319. DOI: 10.1007/s10340-014-0619-x

[39] Epstein D, Gut L, McGhee P, Haas M. Evaluation of new mating disruption formulations. In: Proceedings of the 77th Annual Western Orchard Pest and Disease Management Conference; 15-17 January 2003. Portland, OR: WSU; 2003. p. 8

[40] Waldstein DE, Gut LJ. Effects of rain and sunlight on oriental fruit moth (Lepidoptera: Tortricidae) pheromone microcapsules applied to apple foliage. Journal of Agricultural and Urban Entomology. 2005;**21**:117-128. ISSN (Electronic): 2153-652X

[41] Welter SC, Cave F. Rapid assessment of new pheromone mating disruption devices using an EAG. In: Proceedings of the 79th Annual Western Orchard Pest and Disease Management Conference; 5-7 January 2005. Portland, OR: WSU; 2005. p. 29

[42] Eng JA, Holmes E, Larsen T, Stadlmann S, Ketner K. Effects of sunlight on encapsulated sprayable codling moth pheromone. In: Proceedings of the 77th Annual Western Orchard Pest and Disease Management Conference; 15-17 January 2003. Portland, OR: WSU; 2003. p. 9

[43] Ketner KC. Effect of clay particle film on encapsulated sprayable codling moth pheromone release rate. In: Proceedings of the 76th Annual Western Orchard Pest and Disease Management Conference; 9-11 January 2002. Portland, OR: WSU; 2002. p. 23

[44] Judd GJ, DeLury NC, Gardiner MG. Examining disruption of pheromone communication in *Choristoneura rosaceana* and *Pandemis limitata* using microencapsulated (*Z*)-11-tetradecenyl acetate applied in a laboratory flight tunnel. Entomologia Experimentalis et Applicata. 2005;**114**(1):35-45. DOI: 10.1111/j.0013-8703.2005.00230.x

[45] Wins-Purdy AH, Judd GJR, Evenden ML. Disruption of pheromone communication of *Choristoneura rosaceana* (Lepidoptera: Tortricidae) using microencapsulated sex pheromones formulated with horticultural oil. Environmental Entomology. 2007;**36**:1189-1198. DOI: 10.1603/0046-225X

[46] Trimble RM, Pree DJ, Barszcz ES, Carter NJ. Comparison of a sprayable pheromone formulation and two hand-applied pheromone dispensers for use in the integrated control of oriental fruit moth (Lepidoptera: Tortricidae). Journal of Economic Entomology. 2004;**97**(2):482-489. DOI: 10.1093/jee/97.2.482

[47] Kovanci OB, Walgenbach JF, Kennedy GG. Evaluation of extended-season mating disruption of the oriental fruit moth *Grapholita molesta* (Busck) (Lep., Tortricidae) in apples. Journal of Applied Entomology. 2004;**128**(9-10):664-669. DOI: 10.1111/j.1439-0418.2004.00906.x

[48] Il'Ichev AL, Stelinski LL, Williams DG, Gut LJ. Sprayable microencapsulated sex pheromone formulation for mating disruption of oriental fruit moth (Lepidoptera: Tortricidae) in Australian peach and pear orchards. Journal of Economic Entomology. 2006;**99**(6):2048-2054. DOI: 10.1603/0022-0493-99.6.2048

[49] Hull LA, Myers C, Krawczyk G. Low-rate frequent applications of sprayable

pheromones for oriental fruit moth and codling moth mating disruption in Pennsylvania. In: Proceedings of the 78th Annual Western Orchard Pest and Disease Management Conference; 14-16 January 2004. Portland, OR: WSU; 2004. p. 5

[50] Knight A, Light D, Pickel C, Kovanci O, Molinari F. Management of oriental fruit moth with ground ULV spray applications of a microencapsulated sex pheromone. IOBC-WPRS Bulletin. 2008;**37**:145-151

[51] Stewart-Leslie J. Comparison of sprayable mating disruption to conventional pesticide program for codling moth management in walnuts. In: Proceedings of the 77th Annual Western Orchard Pest and Disease Management Conference; 15-17 January 2003. Portland, OR: WSU; 2003. p. 9

[52] Singleton M, Cave F, Welter S. Mating disruption of codling moths using sprayable formulations: Trap suppression as a first look. In: Proceedings of the 75th Annual Western Orchard Pest and Disease Management Conference; 10-12 January 2001. Portland, OR: WSU; 2001. pp. 141-142

[53] Valdez B, Brunner J. Codling moth and leafroller mating disruption evaluated using sprayable pheromones. In: Proceedings of the 77th Annual Western Orchard Pest and Disease Management Conference; 15-17 January 2003. Portland, OR: WSU; 2003. p. 7

[54] Gut L, Waldstein D, Haas M, McGhee P. Performance of sprayable mating disruption formulations. In: Proceedings of the 76th Annual Western Orchard Pest and Disease Management Conference; 10-12 January 2001. Portland, OR: WSU; 2001. pp. 147-148

[55] Kahn A, Brunner J, Doerr M. Commercial-scale trials of new mating disruption products, 2006. In: Proceedings of the 81st Annual Western Orchard Pest and Disease Management Conference; 10-12 January 2007. Portland, OR: WSU; 2007. p. 45

[56] Knight AL. ULV microencapsulated sex pheromones for codling moth. 2007. Available from: https://treefruitresearch.org/wp-content/uploads/2019/11/Report-420.-KnightSexPheromones.pdf [Accessed: April 25, 2025]

[57] Brunner JF. Evaluation of sprayable pheromone and attract & kill for codling moth and leafroller control. Available from: https://treefrtuitresearch.org/wp-content/uploads/2019/11/Report-091-Brunnerpheromone.pdf [Accessed: April 25, 2025]

[58] Knight AL, Preti M, Basoalto E. What can we learn from dissecting tortricid females about the efficacy of mating disruption programs? Insects. 2025;**16**(3):248. DOI: 10.3390/insects16030248

[59] Knight AL. Targeting *Cydia pomonella* (L.) (Lepidoptera: Tortricidae) adults with low-volume applications of insecticides alone and in combination with sex pheromone. Pest Management Science. 2010;**66**:709-717. DOI: 10.1002/ps.1932

[60] Kovanci OB, Walgenbach JF, Kennedy GC, Schal C. Effects of application rate and interval on the efficacy of sprayable pheromone for mating disruption of the oriental fruit moth *Grapholita molesta*. Phytoparasitica. 2005;**33**:334-342. DOI: 10.1007/BF02981299

[61] Stelinski LL, Gut LJ, Ketner KC, Miller JR. Orientational disruption of codling moth, *Cydia pomonella* (L.) (Lep., Tortricidae), by concentrated formulations of

microencapsulated pheromone in flight tunnel assays. Journal of Applied Entomology. 2005;**129**:481-488. DOI: 10.1111/j.1439-0418.2005.01005.x

[62] Desauziers V, Sicre E, Vignau-Laulhere J, Bourrigaud S, Plaisance H. Suspension of pheromone microcapsules on vine leaves acting as passive dispensers against pests. Environmental Science and Pollution Research. 2022;**29**:14975-14986. DOI: 10.1007/s11356-021-16878-9

[63] Light DM, Knight AL. Microencapsulated pear ester enhances insecticide efficacy in walnuts for codling moth (Lepidoptera; Tortricidae) and navel orangeworm (Lepidoptera: Pyralida). Journal of Economic Entomology. 2011;**104**(4):1309-1315. DOI: 10.1603/EC11058

[64] Knight AL, Light DM, Judd GJ, Witzgall P. Pear ester - From discovery to delivery for improved codling moth management. In: Beck J, editor. Roles of Natural Products for Biorational Pesticides in Agriculture. Washington, DC: ACS Publications; 2018. 83-113 p. DOI: 10.1021/bk-2018-1294.ch008

[65] Kutinkova K, Dzhuvinov V, Lingren B. Control of codling moth, *Cydia pomonella* with Cidetrak® CM-MEC and Cidetrak-CM plus Cidetrak DAMEC liquid formulation in Bulgaria. Available from: www.Ecofruit.net/wp-content/uploads/2020/04/22-kutinkova-124-128pdf [Accessed: April 25, 2025]

[66] Knight AL, Preti M, Basoalto E. Factors impacting the use of an allelochemical lure in pome fruit for *Cydia pomonella* (L.) monitoring. Insects. 2025;**16**:172. DOI: 10.3390/insects16020172

[67] Yee WL, Chapman PS. Effects of GF-120 fruit fly bait concentrations on attraction, feeding, mortality, and control of *Rhagoletis indifferens* (Diptera: Tephritidae).2005. Journal of Economic Entomology. 2005;**98**(5):1654-1663. DOI: 10.1093/jee/98.5.1654

[68] Analytical Software. Statistix 9. In: Analytical Software. FL: Tallahassee; 2003

Chapter 4

Novel IPM Tools for Twenty-First Century Tree Fruits

Alan Lee Knight, Michele Preti and Esteban Basoalto

Abstract

Our review draws attention to a suite of tactics that form the integrated pest management (IPM) approach being developed for use in apple, pear, and cherry to solve new and continuing pest problems. Tactics include development of cultural practices, behavioral tools, spray programs, and biocontrol. All of these have to be embedded in seasonal programs heavily dependent on insecticide sprays. Some approaches are in conflict with sprays, and adjustments have to be made. Many actions support current programs and hopefully will make pest management more integrated with less reliance on pesticides. Others are new tools that can be adopted to gain knowledge and to guide more rational decision-making. The twenty-first century is nearly a quarter done, and our review suggests we have made good progress and that there remains much to do.

Keywords: integrated pest management, apple, pear, cherry, insect pest monitoring

1. Introduction

Pome and stone fruits are tremendously diverse ecosystems available for arthropods colonization including rootzones, trunk and branches, foliage, and fruits. Irrigated orchards planted in dry arid climates present a huge chemical and visual signal to attract mobile pests. Arthropod pests can reduce the vitality of trees by damaging roots or leaves and directly damage the crop by feeding on the shoots, buds, and fruit. Other arthropods fill a secondary trophic level and utilize the pests as resources; thus, they are considered as beneficials. Many of the arthropod species collected in orchards have no impact on crop production, while a few are both pests and beneficials.

Integrated pest management (IPM) draws together accurate data and predictive models to minimize crop injury and consequently economic loss. Many IPM tools developed in the twentieth Century allow these crops to be effectively managed today. Yet new pests arise; some important pests become much less important, and some remain the key pests, driving the entire management program. Climate change together with globalization and limitations in the use of some active ingredients are featuring the present and future concepts and challenges of sustainable fruit production across countries.

Our review focuses on new technologies that have or may allow managers to better seasonally track pests and use new techniques that can manage pest levels and reduce the historical reliance on pesticides in apple, *Malus domestica* (Borkhausen), pear, *Pyrus communis* (L.), and cherry, *Prunus avium* (L.). The economics of crop production is complicated and turbulent based on repeating periods of oversupply and reductions in demand, the opening and closing of international markets, increasing restrictive regulations, and constant evolution in consumer preferences. Our review will present the research that is developing new technologies without due consideration of the economic component to farming success. For this myopic focus, we deeply apologize to the growers!

2. Traditional monitoring

Typically, human scouts walk or drive all-terrain vehicles (ATV's) through orchards to visually sample any number of problems, including the balances between pests, natural enemies, and the crop. Density thresholds have been developed for many of these pests and natural enemies that trigger specific control measures [1]. Orchard scouts may be armed with a hand lens, beating tray, or sweep net. Samples are often collected and examined in the laboratory using leaf brushing machines, or extraction of specimens with water or solvent washing, or Berlese funnels, and the use of microscopes to process the samples. Scouts in the field also check a variety of traps. Male counts from sticky traps baited with sex pheromone lures for moth pests are widely used to support decisions on when or whether to apply insecticides. Sticky cards are also used as traps to monitor smaller pests (scales, mealybugs, thrips, and aphids) and their natural enemies to ascertain pest densities and estimate the potential of biocontrol to avoid unnecessary sprays.

3. New monitoring tools

3.1 Dual sex lures

Dual sex lures have been developed for key tortricid pests attacking tree fruits. These lures have been more effective than sex pheromone lures in monitoring orchards treated with sex pheromone-based mating disruption (MD) and can attract both males and females moth [2, 3]. Knowledge of female moth density and their mating status could be very important to improve MD [4]. Improved phenology models and action thresholds for codling moth (CM), *Cydia pomonella* (L.), have been developed with pear ester, ethyl (*E,Z*)-2,4-decadienoate (PE) [5, 6]. Studies using PE-based lures have found that important information is gained by dissecting female CM [7, 8]. Unfortunately, most pest managers do not sex trapped moths thus, comparing the mating success of females under different types and combinations of MD (hand-applied dispensers, sprayables, and aerosols) has not been possible. Similarly, MD of the other important tortricid pests have not yet been examined using dual sex lures, except for few cases. For instance, the mating status of the European grape vine moth, *Lobesia botrana* (Denis and Schiffermüller), in grapes was shown using the dual sex lure, 2-phenylethanol combined with acetic acid, to be lower in more isolated vineyards and earlier in the season likely due to immigration [9].

The addition of a solar-powered UV-A LED with these dual sex lures can boost the catch of both CM sexes 2 to 4-fold [10, 11]. The UV light pulls inside the moths that have approached the trap but have not yet entered. The use of a single UV-A LED placed inside a delta trap has a limited drawing range and catches few nontarget species in orchards. Using a female moth removal strategy with this lure plus light combination within MD-treated orchards provides a dual approach to manage CM [8, 12].

3.2 Remote monitoring

Remote monitoring is being developed to allow pest populations to be tracked without scouts visiting orchards [13]. Sex pheromone-baited camera traps for CM can correctly identify >90% of the males due to the high specificity of the lure and are gradually being adopted [14]. The use of a camera-equipped traps to monitor multiple pests would likely further reduce monitoring costs. However, the use of camera-equipped traps baited with nonselective kairomone-based lures can be compromised due to excessive false positive identification, especially if multiple pests with similar sizes, silhouettes, and color patterns are captured in the same trap [15].

Further improvements in the algorithms and cameras deployed in camera-equipped traps may be able to reduce these errors [13]. One possible solution to separate species could be counting moths over more limited time frames during the peak flight period of each species. Technical modifications will be needed and may require enhanced solar and battery power to take more photographs per day. A second promising approach could be to use differences in wing vibrational frequencies to sort insect species [16].

3.3 Host plant volatile lures

Host plant volatile lures have been developed to both monitor and pull biocontrol agents into orchards [17, 18]. Traps with a four-component lure (acidic acid, methyl salicylate, 2-phenylethanol, and phenylacetaldehyde) outperformed beat trays, and tracked 12 natural enemies within pear, including some not caught on the trays [19]. The replacement of beat trays with host plant volatile-baited traps also significantly reduced labor costs for monitoring. The use of these traps is allowing action thresholds to be developed for natural enemies and helps to optimize spray timings [20].

3.4 Canines'

Canines' innate ability to smell is being developed to detect several types of cryptic orchard pests. Dogs have been used to provide detections of early symptoms of a number of phytoplasmas in several crops [21], including for the Western X Virus in cherry, a disease that has caused huge losses in Washington State [22]. Early detection of phytoplasmas is difficult yet essential for deploying effective strategies. Visual sampling and molecular techniques are too inefficient or expensive to be widely used. In addition, canines have been trained to locate extra orchard overwintering populations of the brown marmorated stink bug (BMSB), *Halyomorpha halys* (Stål) [23], and the recently introduced fulgorid, spotted lanternfly, *Lycorma delicatula* (White) [24].

3.5 Biotremology

Biotremology, the study of substrate-borne mechanical waves in animal communication, has emerged as a promising tool for various heteropteran pests of tree fruits including leafhoppers, stink bugs, and psyllids [25]. These pest groups rely on vibrational signals for mate recognition and courtship. Artificially generated vibrations can interfere with these signals, preventing successful mating [26]. Vibrational traps that exploit insects' natural communication signals can attract pests to a specific location, where they are either captured or exposed to insecticides, "vibration-based attract-and-kill" strategies [27]. For example, in psyllids, which are key vectors of plant pathogens, vibrational traps emit disruptive mating signals, luring individuals onto a trapping surface and reducing population densities [28]. The use of vibrations to disrupt mating in pear psylla (PP), *Cacopsylla pyricola* (Foerster), is being developed [29]. Biotremology can also enhance biocontrol by using the vibrational sensitivity of natural enemies. For instance, studies on whiteflies and aphids suggest that exposing parasitoids to host-generated vibrations before their release can enhance their ability to locate prey in the field, thereby increasing their efficacy in suppressing pest populations [30].

4. Traditional IPM programs

Traditional IPM programs in tree fruits are widely considered to be successful [31], but there are many pests to manage, and insecticide use remains the most effective and dependable tools to apply [32]. The twenty-first century has brought insecticides with new mode-of-actions and ones that are often more selective than the chemistries relied on in the last Century [33, 34]. Without any new broad spectrum "silver bullets" available, research has focused on implementing new even more integrated programs [35]. Unfortunately, invasive pests pose fresh disturbances and fractures in these programs [36].

5. New IPM programs

New IPM programs in apple, pear, and cherry are more unstable than many practitioners hope. Several problems are on the horizon and could continue to rise. Selection for pesticide resistance has always been a determinant factor shaping management programs of key pests in tree fruits, that is, CM and PP [37, 38]. The diamide chlorantraniliprole is the most common insecticide used for CM and apparently has lost its efficacy in the mid-Atlantic region of the USA [39]. Organic apple growers in Washington State have reported some resistance to the granulosis virus [40]. New pests generally require applications of more broad-spectrum insecticides to achieve control, at least initially, until IPM programs can be developed and stabilized [41–43].

Optimists may believe that the development of new tools to track pests and crops and automated spray and tractor-based precision technologies are revolutionary advances that can improve crop production [44]. Surely, they provide some refinement in management, but they do not eliminate the problems associated with farming large monocultures. Poor economics often inversely drives the quality of the pest management applied. In tough years, the more you grow, the more you lose! Even in good years, margins are thin for most growers. The transfer of family to corporate farming is speeding up. Again, optimists think that larger, well-funded operations will use the best practices and be the most successful [44]. Yet the consolidation of farming into industrial level production units tends to further reduce ecosystem diversity and increase their fragility [45]. Unless research can build better IPM, the twenty-first century will surely be interesting, and the insects will be there to testify!

5.1 "Old Friends Never Die!"

"Old Friends Never Die!" Each of the three crops in this review retains key pests that require effective management. CM in both apple and pear orchards; PP in pear; and temperate fruit flies in cherry, across orchards in both North America and Europe. Here, we highlight some new tools that will likely impact each of their management programs.

5.2 Mating disruption for CM

Mating disruption for CM and the key tortricid pests in tree fruits has continued to grow and is estimated at >1000,000 ha [46]. Studies with CM have found that levels of unmated females can be low despite the use of MD [8]. Efforts to improve CM MD can likely be achieved with increases in the rate of MD, such as combing hand-applied dispensers with either sprayables or aerosols [47].

Aerosol formulations of sex pheromone were first developed in 1994 [48]. One factor driving the adoption of this technology was the reduced labor required for deployment of 2.5 units ha^{-1} compared with up to 1000 hand-applied dispensers ha^{-1}. However, initial concerns about its use in small irregular-shaped blocks limited its adoption [49]. Today, aerosols have become the fastest growing MD technology in fruit and nut crops [46]. Research studies have created a lower cost program (50% lower) with aerosols for CM. Internal grids of aerosols placed at 1 ha^{-1} with 10-m border hand-applied dispenser applications (1000 ha^{-1}) were effective for CM in small and large blocks [49]. The rate and frequency of the aerosol sprays also could be reduced substantially to further reduce the cost [50]. Unfortunately, manufacturers have not wanted to reduce these rates.

Pear ester is used in combination with sex pheromone in several MD dispenser types for CM and in microencapsulated formulations [51] and provides greater disruption of female mating [4, 52]. However, aerosol formulations of pear ester have not yet been tested. Another innovative use of CM MD not yet adopted is low-volume sprays of a microencapsulated formulation, which were effective and provided growers with more flexibility in using MD for key periods of pest abundance [53].

5.3 Sterile insect releases of CM

Sterile insect releases of CM has been conducted in Canada since 1992 [54]. The public-funded program has evolved from an ill-conceived eradication effort to a very expensive program for only one pest (estimated at $ 1120 ha^{-1}) as tree fruit production has declined 50% in the region. Unfortunately, new pests have emerged under the Canadian CM Sterile Insect Technique (SIT) program, such as the tortricid, *Spilonota ocellana* (Denis and Schiffermuller), and the sesiid, *Synanthedon myopaeformis* (Borkhausen) [55, 56].

The sterilized CM have been distributed to several areas. For example, moths from 1995 to 1999 were transported to the USA to treat nearly 400 ha adjacent to the two countries border in an area-wide program combined with MD that was a very successful partially funded government program [57]. Moths produced after the field season in the northern hemisphere have also been shipped to South Africa and New

Zealand for smaller pilot projects [58, 59]. Beginning in 2018, the excess production of Canadian sterile CM (ca 2,000 ha) was made available to growers in the USA and has been commercially released into individual orchards *via* drone. A three-year funded program examined the effectiveness of SIT as a component of organic management including MD and sprays in small individual orchards of 2–3 ha [60]. The program did not reduce levels of injury compared with the orchards not receiving moths. Some growers believe SIT is a useful approach to reduce high levels of CM, but it takes several years. Further expansion of this approach will require the development of additional rearing capacity likely in the USA. The use of X-rays to sterilize CM may facilitate further adoption of this technology [61].

5.4 Granulosis virus of CM

Granulosis virus of CM has been a key tactic used primarily in organic production [62]. Extremely high levels of resistance developed in Europe [63], and limited resistance has been reported in North America [64]. New products using other more effective virus strains have been developed [65]. The virus is typically used with oil and a microencapsulated pear ester formulation [66].

5.5 Pear psylla IPM

Pear psylla IPM is focused on managing both *C. pyricola* in North America and *C. pyri* (L.) in Europe [67]. Psyllid feeding and their production of honeydew causes russeting and black mold growth on the fruit and a downgrade to a processing market. Conventional pear management can include >10 sprays annually for psyllids and nearly 50% of the production costs [68]. Biocontrol of psyllids in these orchards is ineffective. In addition, PP has a remarkable history in developing resistance to insecticides, which limit the available chemical tools [38].

Several generalist predators can regulate PP populations in pear, such as mirids, lacewings, and earwigs [69]. Unfortunately, when and whether these agents will arrive at orchards and at a sufficient density to effectively crash PP populations is variable. Thresholds have been established for the key natural enemies [20]. Monitoring of key natural enemies with novel traps has replaced beating trays and allows managers to better assess their populations [70].

Many pear growers are not able to hold back from spraying while waiting for the uncertainty of effective biocontrol of PP to establish [68]. In some cases, selective integrated programs often become more effective after one or more years of adoption [71]. Thus, growers must face a management "tightrope" every year to deal with PP. Integrated studies have created successful program strategies including selective insecticides, particle films, fruit washing, and reflective ground covers, but growers must persevere with these new integrated tactics to be effective [71].

5.6 Cherry

Cherry pest management is now impacted by two to three instead of one key pest. The management programs were based primarily on the tephritids *Rhagolitis cerasi* (L.) and *R. indifferens* (Curran), in Europe and North America, respectively, but are now upended by the spotted-wing drosophila (SWD), *Drosophila suzukii* (Matsumura), and outbreaks of *Candidatus Phytoplasma pruni* vectored by several leafhopper species [41, 43, 72]. Development of an organic bait (GF-120) for the *Rhagolitis* spp. is not effective for SWD [73]. Beyond the increase spray applications to manage these new pests, reflective ground cover, management of ground cover plants, and post-harvest particle films have been tested for leafhoppers [74]. New tools for SWD will be reviewed in a later section.

6. Expanding old technologies

Expanding old technologies in tree fruits include several approaches that can be used for more than one pest.

6.1 Particle film technology

Particle film technology utilizes chemically inert clay or minerals applied over the tree canopy to protect crops from insects and heat stress [75]. The most commonly used particle films are kaolin clays, diatomaceous earth, and calcium carbonate powders. Arthropods such as psyllids and tortricid larvae can be repelled/deterred from settling or feeding on leaves and fruit coated with particle films [76, 77]. Particle films can have both positive and negative effects on orchard pests with their disruption of biocontrol, considered to be the greatest problem with repetitive use [78]. Currently, kaolin is used extensively pre-bloom in pears to reduce the oviposition of PP [71]. Summer use of particle films has been discouraged. However, the addition of 1% horticultural oil with particle films helps to suppress outbreaks of aphids. The impact of the extended summer use of sunburn protection with various particle films on pest management has surprisingly not been examined.

6.2 Pest exclusion

Pest Exclusion whether achieved at an orchard or a landscape level can provide significant pest protection. Immigration of CM females is a key factor influencing orchard management success [79]. Area-wide management programs where neighboring growers work together to clean up CM infestations over a large area have been very successful [57, 80].

Exclusion netting provides a similar approach, but for individual growers and at a farm scale. The use of netting in apple and cherry orchards serves several important management purposes, including protection against hail, UV radiation, frugivorous birds, and insect pests [81]. The adoption of overhead permanent structures in the arid region of eastern Washington State has been primarily for sunburn protection of fruit [82]. Exclusion netting in Europe has been more directed for CM due to insecticide resistance developing to both organic and conventional insecticides [63, 83]. Exclusion techniques evolved into single-row netting (Alt'Carpo) that completely enclosed each row separately and was removed each year [84]. Single-row netting was shown to be more effective against CM than the traditional hail netting [85]. The effectiveness of netting for CM appears to be placing netting tight around the canopy to interfere with moth sexual behaviors [86].

The positive impact of exclusion netting on several orchard pests has been detailed [56, 81]. Conversely, a few pests have become a more significant problem under nets [81, 87, 88]. The use of netting can also reduce levels of avian-caused fruit injury in certain varieties, such as "Honeycrisp" [86].

6.3 Biological control

Biological control has become a much more important factor considered in constructing sustainable IPM programs. Organic spray programs are clearly crafted with some intention of maximizing biocontrol of both key and secondary pests [89]. Establishing natural enemies in conventional orchards is more difficult due to the seasonal application of pesticides. Classical and augmentative biocontrol are both important and have focused on both newly introduced pests such as BMSB and SWD and classic pests such as PP, aphids, and mealybugs.

Organic production of tree fruits has continued to increase since 2000, but oversupply has reduced the price differential it previously enjoyed over conventional fruit [90]. Effective organic practices have continued to be limited, and some pests such as CM can be difficult to manage [91]. Other pests tend to be less important under organic management due to biocontrol, such as tortricid leafrollers [87, 89].

Generalist predators of many pests can play an important role in tree fruits, and their numbers and timing of their arrival are key impact factors [89]. Phenology models have been developed to optimize release timings and sprays [20, 92]. Landscape ecology is a determinant factor for the level of biocontrol attained in orchards [93]. Growers can also supplement biocontrol by purchasing natural enemies from insectaries and releasing by hand or with drones [94]. Efforts to draw natural enemies into the orchard have included the use of host plant volatile lures [17]. Augmentative transfers of the earwig, *Forficula auricularia* (L.), from stone fruit where they are considered to be a pest [95], to pear has been attempted [96]. Yet establishing earwigs is considered likely only if the orchard is transitioning to organic practices [96]. Thus, without organic production practices or a reduced conventional (i.e., IPM) spray program, realizing benefits from natural enemies in tree fruits is a struggle [97].

6.4 MD for non-lepidopteran pests

MD for non-lepidopteran pests now includes both coccid scale and pseudococcid mealybug pests (Hemiptera: Coccoidea) [98]. Within tree fruits, sex pheromone has been tested in successful MD field trials for San José Scale, *Comstockaspis perniciosa* (Comstock) [99]. The sex pheromone for pear psyllids have been identified but have not been tested in MD trials [100, 101]. The complex syntheses and formulations required with these pests are the major limitation in fully evaluating their potential. Also, the landscape ecology of the psyllids in and out of orchards likely limits this approach [102].

7. The globalization of new pests (a plague upon all our houses)

The globalization of new pests (a plague upon all our houses) continues to be a threat to sustaining IPM in tree fruits. The loss of established selective programs relying on biocontrol and minimal sprays typically occurs when a new pest arrives in the orchard. For example, organic cherry production has been restricted by SWD [41]. The use of MD for CM is reconsidered if BSMB needs to be managed [42]. Also, new endemic pests have become more important, that is, Western X virus in cherry requiring extra sprays to kill leafhoppers that were not previously pests [43].

7.1 BMSB

BMSB is a classic introduced-pest disrupting the established crop management that requires the development of area-wide "all available tools" IPM programs to achieve success [103]. Unfortunately, BMSB can feed on >300 plant species, and adults and nymphs move into orchards throughout the season from the surrounding landscape [104]. Early detection and novel management approaches to avoid the sole reliance on disruptive insecticide-based programs in tree fruits are essential. An aggregation pheromone is used in combination with a synergistic compound to monitor BMSB [42]. Monitoring is typically with sticky panels or pyramid-shaped traps [105]. An action threshold has been defined [106], and area-wide pheromone-based trap networks have been deployed for monitoring [107]. Traps with both visual and chemical stimuli have been developed [108, 109]. BMSB adults communicate with vibrations, and studies have used these signals in combination with the pheromone to improve trapping [110]. Automated pest detection for BMSB is under investigation with different approaches, including the use of smart traps [111] and unmanned aerial vehicles [112].

BMSB management is primarily based on the chemical control with insecticides [113]. In orchards, the switch back to broad-spectrum insecticides and the need for repeated interventions across the season have disrupted the IPM programs for other pests [42]. New approaches have been tested for BSMB to minimize this disruption, such as weekly alternate row middle [106] and border sprays [114]. Noninsecticidal tools to manage BSMB include a number of simultaneous approaches. These can include: (*i*) exclusion netting [115]; (*ii*) natural products, kaolin, diatomaceous earth, sulfur [116]; (*iii*) attract-and-kill and mass trapping with its aggregation pheromone with or without insecticides using nets placed surrounding orchards [117–119]; (*iv*) trap cropping with alternative hosts [120]; and (*v*) natural and classical biocontrol with egg parasitoids [121, 122].

7.2 SWD

SWD is a destructive pest of small fruits including cherry [41]. Native to Western Asia, SWD has exhibited a rapid global expansion since 2008 [41]. Various trapping and monitoring strategies have been developed to detect and manage SWD populations [123]. Unfortunately, sex pheromone extracts have proven ineffective for trapping SWD [124]. Recent development of sticky red cards with synthetic lures has greatly improved monitoring [125]. Rapid molecular assays have been developed to identify SWD [126]. Automatic detection of SWD among similar species has been developed using optical identification of wingbeats [127]. Deep learning algorithms have been programmed to sex SWD adults for use with a potential SIT program [128].

Effective management of SWD requires an integrated approach combining cultural, biological, and chemical control methods. Rapid harvest schedules, growing early-season fruit, removing dropped fruits after harvest, and pruning trees to improve sunlight penetration are effective nonchemical approaches [129]. Exclusion netting of cherry with a fine mesh provides a physical barrier against SWD infestation [130]. Parasitoid wasps have shown promise in reducing SWD populations, though their effectiveness varies by region [131, 132].

Insecticides remain a primary control method for SWD in cherry orchards, and resistance management involves rotating classes of active ingredients [133]. Insecticide efficacy can be enhanced by adjuvants that improve coverage and retention, as well as feeding stimulants that increase ingestion [73, 134, 135]. Attract-and-kill (infect) tactics and mass trapping techniques incorporating insecticides have been developed [136–138]. Area-wide control strategies with SIT and genetic control have been proposed [139, 140].

8. Conclusions

The human population is still exploding on earth, and agriculture faces a constant demand for greater output while maintaining excellent quality. Governmental regulations that restrict tools that enhance production in order to "protect" public health are in direct conflict with this population growth. We believe as entomologists that technology can serve to create both safer and more abundant agricultural production. However, it will require increased funding to support the research needed to develop novel and more effective IPM tools.

Author details

Alan Lee Knight[1], Michele Preti[2*] and Esteban Basoalto[3]

1 Instar Biologicals, Yakima, Washington, USA

2 Integrated Pest Management, Faenza, Italy

3 Facultad de Ciencias Agrarias y Alimentarias, Universidad Austral de Chile, Valdivia, Chile

*Address all correspondence to: mpreti1990@gmail.com

References

[1] Beers EH, Hull LA, Jones VP. Sampling pest and beneficial arthropods of apple. In: Pedigo LP, Buntin GD, editors. Handbook of Sampling Methods for Arthropods in Agriculture. 1st ed. Boca Raton: CRC Press; 1993. 34 p. DOI: 10.1201/9781003067900

[2] Knight AL, El-Sayed AM, Judd GJR. Development of 2-phenylethanol plus acetic acid lures to monitor oblique banded leafroller (Lepidoptera: Tortricidae) under mating disruption. Journal of Applied Entomology. 2017;**141**:729-739. DOI: 10.1111/jen.12393

[3] Knight AL, Preti M, Basoalto E. Factors impacting the use of an allelochemical lure in pome fruit for *Cydia pomonella* (L.) monitoring. Insects. 2025a;**16**:172. DOI: 10.3390/insects16020172

[4] Knight AL, Preti M, Basoalto E. What can we learn from dissecting tortricid females about the efficacy of mating disruption programs? Insects. 2025b;**16**(3):248. DOI: 10.3390/insects16030248

[5] Knight AL, Light DM. Timing of egg hatch by early-season codling moth (Lepidoptera: Tortricidae) predicted by moth catch in pear ester- and codlemone-baited traps. Canadian Entomologist. 2005a;**137**:728-738. DOI: 10.4039/n05-039

[6] Knight AL, Light DM. Developing action thresholds for codling moth (Lepidoptera: Tortricidae) with pear ester- and codlemone-baited traps in apple orchards treated with sex pheromone mating disruption. Canadian Entomologist. 2005b;**137**:739-747. DOI: 10.4039/n05-040

[7] Light DM, Grant JA, Haff RP, Knight AL. Addition of pear ester with sex pheromone enhances disruption of mating by female codling moth (Lepidoptera: Tortricidae) in walnut orchards treated with meso dispenser. Environmental Entomology. 2017;**46**:319-327. DOI: 10.1093/ee/nvw168

[8] Knight AL, Preti M, Basoalto E, Mujica VM, Favaro R, Angeli S. Combining female removal with mating disruption for management of *Cydia pomonella* in apple. Entomologia Generalis. 2021;**42**:309-321. DOI: 10.1127/entomologia/2021/1316

[9] Larsson-Herrera S, Rikk P, Köblös G, Szelényi MO, Molnár BP, Dekker T, et al. Designing a species-selective lure based on microbial volatiles to target *Lobesia botrana*. Scientific Reports. 2020;**10**:6512. DOI: 10.1038/s41598-020-63088-3

[10] Knight AL, Preti M, Basoalto E, Fuentes-Contreras E. Increasing catches of adult moth pests (Lepidoptera: Tortricidae) in pome fruit with low-intensity LED lights added to sex pheromone/kairomone lure-baited traps. Journal of Applied Entomology. 2023;**147**:843-856. DOI: 10.1111/jen.13176

[11] Knight AL, Mujica V, Basoalto E, Preti M. Simultaneous effective monitoring of *Grapholita molesta* and *Cydia pomonella* (Lepidoptera: Tortricidae) in traps with a dual sex pheromone/kairomone lure plus a UV-A light. Journal of Applied Entomology. 2024;**148**:1261-1275. DOI: 10.1111/jen.13344

[12] Preti M, Knight AL, Mujica MV, Basoalto E, Favaro R, Angeli S. Developing female removal for *Cydia*

pomonella (Lepidoptera: Tortricidae) in organic pear in the USA and Italy. Journal of Applied Entomology. 2021a;**145**:856-868. DOI: 10.1111/jen.12918

[13] Preti M, Verheggen F, Angeli S. Insect pest monitoring with camera-equipped traps: Strengths and limitations. Journal of Pest Science. 2021b;**94**:203-217. DOI: 10.1007/s10340-020-01309-4

[14] Suto J. Codling moth monitoring with camera-equipped automatic traps: A review. Agriculture. 2022;**12**:1721. DOI: 10.3390/agriculture12101721

[15] Preti M, Favaro R, Knight AL, Angeli S. Remote monitoring of *Cydia pomonella* adults among assemblage of nontargets in sex pheromone-kairomone-baited smart traps. Pest Management Science. 2021c;**77**:4084-4090. DOI: 10.1002/ps.6433

[16] Burks CS. Comparison of navel orangeworm adults detected with optical sensors and captured with conventional sticky traps. AgriEngineering. 2022;**4**:523-532. DOI: 10.3390/agriengineering4020035

[17] Mills NJ, Jones VP, Baker CC, Melton TD, Steffan SA, Unruh TR, et al. Using plant volatile traps to estimate the diversity of natural enemy communities in orchard ecosystems. Biological Control. 2016;**102**:66-76. DOI: 10.1016/j.biocontr.2016.05.001

[18] Ayelo PM, Pirk CWW, Yusuf AA, Chailleux A, Mohamed SA, Deletre E. Exploring the kairomone-based foraging behaviour of natural enemies to enhance biological control: A review. Frontiers in Ecology and Evolution. 2021;**9**:641974. DOI: 10.3389/fevo.2021.641974

[19] Adams C, Schmidt-Jefferies R, Orpet R. Calibrating Current NE Action Thresholds with Lure Baited Trap Catch [Internet]. 2024. Available from: https://treefruitresearch.org/report/calibrating-current-ne-action-thresholds-with-lure-baited-trap-catch/ [Accessed: February 20, 2025]

[20] DuPont ST, Strohm CJ, Kogan C, Hilton R, Nottingham L, Orpet R. Pear psylla and natural enemy thresholds for successful integrated pest management in pears. Journal of Economic Entomology. 2023;**116**:1249-1260. DOI: 10.1093/jee/toad101

[21] Verma A, Verma R, Tripathi S. Scent discriminating dogs for easy detection and surveillance of phytoplasma diseases - A new possibility. Bioteca Research Today. 2022;**4**:834-836

[22] Harper S, Northfield T, Nottingham L, DuPont T. X-Disease Phytoplasma (Western X). Washington State University Extension [Internet]; 2021. Available from: http://treefruit.wsu.edu/crop-protection/disease-management/western-x/ [Accessed: January 15, 2025]

[23] Lee DH, Cullum JP, Anderson JL, Daugherty JL, Beckett LM, Leskey TC. Characterization of overwintering sites of the invasive brown marmorated stink bug in natural landscapes using human surveyors and detector canines. PLoS One. 2014;**9**:e91575. DOI: 10.1371/journal.pone.0091575

[24] Fuller AK, Augustine BC, Cliftoon EH, Beese J, Hurt A, Brown-Lima J. Effectiveness of canine-assisted surveillance for early detection of invasive spotted lanternfly. Ecosphere. 2024;**15**:e70113. DOI: 10.1002/ecs2.70113

[25] Nieri R, Beukeboom LW, Mazzoni V. Insect biotremology - An introduction. Entomologia Experimentalis et Applicata. 2024;**172**:1113-1115. DOI: 10.1111/eea.13520

[26] Polajnar J, Maistrello L, Bertarella A, Mazzoni V. Vibrational communication of the brown marmorated stink bug (*Halyomorpha halys*). Physiological Entomology. 2016;**41**:249-259. DOI: 10.1111/phen.12150

[27] Sönmez C, Özgen I. New trend in pest management: Biotremology. Journal of Entomology and Zoology Studies. 2025;**13**:14-19. DOI: 10.22271/j.ento.2025.v13.i1a.9437

[28] Pekas A, Mazzoni V, Appel H, Cocroft R, Dicke M. Plant protection and biotremology: Fundamental and applied aspects. Trends in Plant Science. 2024;**29**:32-39. DOI: 10.1016/j.tplants.2003.06.021

[29] Jocson DMI, Gonzales MT, Horton DR, Nottingham LB, Beers EH, Oeller LC, et al. Characterizing substrate-borne vibrational mating signals produced by pear psylla, *Cacopsylla pyricola* (Hemiptera: Psyllida). Journal of Insect Behavior. 2023;**36**:267-276. DOI: 10.1007/s10905-023-09839-2

[30] Zippari C, Nieri R, Verrastro V, Mandrioli M, Grasso DA, Mazzoni V, et al. Exploring the role of vibrational cues in a parasitoid-host interaction. In: Proceedings of the 3rd Biotremology Conference; 19-22 September 2022; Piran, Slovenia. p. 9. Available from: https://hdl.handle.net/10449/78415

[31] Damos P, Colomar LAE, Ioratti C. Integrated fruit production and pest management in Europe: The apple case study and how far we are from the original concept? Insects. 2015;**6**:626-657. DOI: 10.3390/insects6030626

[32] Chouinard G, Pelletier F, Vincent C. Pest activity and protection practices: Four decades of transformation in Quebec apple orchards. Insects. 2021;**12**:197. DOI: 10.3390/insects12030197

[33] Beers EH, Horton DR, Miliczky E. Pesticides used against *Cydia pomonella* disrupt biological control of secondary pests of apple. Biological Control. 2016;**102**:35-43. DOI: 10.1016/j.biocontrol.2016.05.009

[34] Radrigán-Navarro C, Beers EH, Alvear A, Fuentes-Contreras E. Acute toxicity of lethal and sublethal concentrations of neonicotinoid, insect growth regulator and diamide insecticides on natural enemies of the woolly apple aphid and the obscure mealybug. Chilean Journal of Agricultural Research. 2021;*81*(3):398-407. DOI: 10.4067/S0718-58392021000300398

[35] Barrera JF. Beyond IPM: Introduction to the Theory of Holistic Pest Management. NY: Springer International Publishing; 2020. 154 p. DOI: 10.1007/978-3-030-43370-3

[36] Peterson RKD, Higley LG, Pedigo LP. Whatever happened to IPM? American Entomologist. 2018;**64**(3):146-150. DOI: 10.1093/ae/tmy049

[37] Kadoić Balaško M, Bažok R, Mikac KM, Lemic D, Pajač Živković I. Pest management challenges and control practices in codling moth: A review. Insects. 2020;**11**:38. DOI: 10.3390/insects11010038

[38] Civolani S. The past and present of pear protection against the pear psylla, *Cacopsylla pyri* L. In: Perveen F, editor. Insecticides - Pest Engineering. London, UK: Intech Open; 2012. pp. 385-408. DOI: 10.5772/1354

[39] Krawczyk G, Mellott L, Gere J, Seutter T. Developing challenges in managing codling moth with diamide insecticides. In: Proceedings of the Oregon Pest and Disease Management Conference 99; 15-17 January 2025; Portland. OR: OPDMC; 2025. p. 57

[40] Nielsen AL, Rucker A. Update on the status of resistance to CpGV in Washington codling moth. In: Proceedings of the Oregon Pest and Disease Management Conference 99; 15-17 January 2025; Portland. OR: OPDMC; 2025. p. 39

[41] Beers E, Van Steenwyk R, Shearer P, Coates W, Grant J. Developing *Drosophila suzukii* management programs for sweet cherry in the western United States. Pest Management Science. 2011;**67**:1386-1395. DOI: 10.1002/ps.2279

[42] Leskey TC, Short BD, Ludwick D. Comparison and refinement of integrated pest management tactics for *Halyomorpha halys* (Hemiptera: Pentatomidae) management in apple orchards. Journal of Economic Entomology. 2020;**113**:1725-1734. DOI: 10.1093/jee/toaa067

[43] Clarke AE, Catron KA, Corral CR, Marshall AT, Adams CG, Cooper WR, et al. *Colladonus* spp. (Hemiptera: Cicadellidae) vectors of X-disease: Biology and management in western United States. Journal of Integrated Pest Management. 2024;**15**:1-10. DOI: 10.1083/jipm/pmae005

[44] Van Lanen AL. The Washington Apple: Orchards and the Development of Industrial Agriculture. Norman, OK: University of OK Press; 2022. p. 298

[45] Daelemans R, Hulsmans E, Fockaert L, Vranken L, De Bruyn L, Honnay O. Agroecosystem multifunctionality of apple orchards in relation to agricultural management and landscape context. Ecological Indicators. 2023;**154**:110496. DOI: 10.1016/j.ecolind.2023.110496

[46] Gut LJ, Adams C, Miller J, McGhee P, Thomson D. Biological control in integrated management of deciduous fruit insect pests: The use of semiochemicals. In: Xu X, Fountain M, editors. Integrated Management of Diseases and Insect Pests of Tree Fruit. Philadelphia: Burleigh Dodds Sci. Publishing; 2019. pp. 571-662. DOI: 10.1201/9780429266690

[47] Knight AL, Preti M, Basoalto E. What can we learn from dissecting tortricid females about the efficacy of mating disruption programs? Insects. 2025c:**16**(3);248. DOI: 10.3390/insects16030248

[48] Shorey HH, Gerber RG. Use of puffers for disruption of sex pheromone communication of codling moths (Lepidoptera; Tortricidae) in walnut orchards. Environmental Entomology. 1996;**25**:1398-1400. DOI: 10.1093/ee/25.6.1398

[49] Knight AL. Development of aerosol devices for management of codling moth and leafrollers. IOBC-WPRS Bulletins. 2002;**25**:101-110

[50] McGhee PS, Miller JR, Thomson DR, Gut LJ. Optimizing aerosol dispensers for mating disruption of codling moth, *Cydia pomonella* L. Journal of Chemical Ecology. 2016;**42**:612-616. DOI: 10.1007/s10886-016-0724-9

[51] Knight AL, Light DM, Judd GJR, Witzgall P. Pear ester–from discovery to delivery for improved codling moth management. In: Beck JJ, Rering CC, Stephen Duke SO, editors. Roles of Natural Products for Biorational Pesticides in Agriculture. Washington, DC: ACS Symposium Series; American Chemical Society; 2018. pp. 83-113. DOI: 10.1021/bk-2018-1294.ch008

[52] Knight AL, Light DM. Combined approaches using sex pheromone and pear ester for behavioral disruption of codling moth (Lepidoptera: Tortricidae). Journal of Applied Entomology. 2014;**138**:96-108. DOI: 10.1111/jen.12071

[53] Knight AL, Larson TL, Ketner K, Hilton R, Hawkins L. Field evaluations of concentrated spray applications of microencapsulated sex pheromone for codling moth (Lepidoptera: Tortricidae). Environmental Entomology. 2008;**37**:980-989. DOI: 10.1093/ee/37.4.980

[54] Thistlewood HMA, Judd GJR. Twenty-five years of research experience with the sterile insect technique and area-wide management of codling moth, *Cydia pomonella* (L.), in Canada. Insects. 2019;**10**(9):292. DOI: 10.3390/insects10090292

[55] McBrien HL, Judd GJR. Emergence of overwintered larvae of eye-spotted bud moth, *Spilonota ocellana* (Lepidoptera: Tortricidae) in relation to temperature and apple tree phenology at Summerland, British Columbia. Journal of the Entomological Society of British Columbia. 2004;**101**:15-23

[56] Nelson TD, Haye T, Richardson T, MacDonald JL, Judd GJR, Moffat CE. *Synanthedon myopaeformis* (Borkhausen), apple clearwing moth/Sésie du pommier (Lepidoptera: Sesiidae). In: Vankosky MA, Martel V, editors. Biological Control Programmes in Canada 2013-2023. Croydon, UK: CABI - CPI Group Ltd; 2024. pp. 385-393. DOI: 10.1079/9781800623279.0042

[57] Knight AL. Codling moth areawide integrated pest management. In: Koul O, Cuperus G, Elliott N, editors. Areawide Pest Management: Theory and Implementation. Wallingford UK: CABI; 2008. pp. 159-190. DOI: 10.1079/9781845933722.0159

[58] Bloem S, Carpenter JE, Blomefield TL, Harrison C. Compatibility of codling moths *Cydia pomonella* (L) (Lep. Tortric.) from South Africa with codling moths shipped from Canada. Journal of Applied Entomology. 2010;**134**:201-206. DOI: 10.1111/j.1439-0418.2009.01416.x

[59] Horner RM, Lo PL, Rogers DJ, JTS W, Suckling DW. Combined effects of mating disruption, insecticides, and the sterile insect technique on *Cydia pomonella* in New Zealand. Insects. 2020;**11**:837. DOI: 10.3390/insects11120837

[60] Beers EH, Crowder D. Optimizing Sterile Insect Release of Codling Moth in Washington. Final report to Washington Tree Fruit Research Commission, Wenatchee, WA [Internet]. 2021. Available from: https://treefruitresearch.org/report/optimizing-sterile-insect-release-of-codling-moth-in-washington/ [Accessed: January 21, 2025]

[61] Moses-Gonzales N, Gall ML, Richter J, Lieberman JB. Enhancing radiological security in sterile insect technique: X-rays for codling moth sterilization. In: Proceedings of the Oregon Pest and Disease Management Conference 99; 15-17 January 2025; Portland. OR: OPDMC; 2025. p. 20

[62] Lacey LA, Thomson D, Vincent C, Arthurs SP. Codling moth granulosis virus: A comprehensive review. Biocontrol Science and Technology. 2008;**18**:639-663. DOI: 10.1080/09583150802267046

[63] Sauphanor B, Berling M, Toubon JF, Reyes M, Delnatte J, Allemoz P. Cases of resistance to granulosis virus in the codling moth. Phytoma. 2006;**590**:24-27

[64] Fan J, Jehle JA, Rucker A, Nielsen AL. First evidence of CpGV resistance of codling moth in the USA. Insects. 2022;**13**:533. DOI: 10.3390/insects13060533

[65] Eberle KE, Asser-Kaiser S, Sayed SM, Nguyen HT, Jehle JA. Overcoming

the resistance of codling moth against conventional *Cydia pomonella* granulosis (CPGV-M) by a new isolate CpGV-112. Journal of Invertebrate Pathology. 2008;**98**:293-298. DOI: 10.1016/j.jip.2008.03.003

[66] Arthurs SP, Hilton R, Knight AL, Lacey LA. Evaluation of the pear ester kairomone as a formulation additive for the granulosis virus of codling moth (Lepidoptera: Tortricidae) in pome fruit. Journal of Economic Entomology. 2007;**100**:702-709. DOI: 10.1093/jee/100.3702

[67] Civolani S, Soroker V, Cooper RW, Horton DR. Diversity, biology and management of the pear psyllids: A global look. Annals of the Entomological Society of America. 2023;**116**:331-357. DOI: 10.1093/aesa/sand025

[68] DuPont ST, Strohm CJ, Nottingham L, Rendon D. Evaluation of an integrated pest management program for central Washington pear orchards. Biological Control. 2021;**152**:104390. DOI: 10.1016/j.biocontrol.2020.104390

[69] DuPont ST, Strohm CJ. Integrated pest management programmes increased natural enemies of pear psylla in central Washington pear orchards. Journal of Applied Entomology. 2020;**144**:109-122. DOI: 10.1111/jen.12694

[70] Horton DR, Miliczky ER, Lewis TM, Wohleb CH, Waters TD, Dickens AA, et al. Building a better Psylloidea (Hemiptera) trap? A field look at a prototype constructed using a three-dimensional printer technology. Canadian Entomologist. 2018;**151**:115-129. DOI: 10.4039/tce.2018.59

[71] Nottingham LB, Orpet RJ, Beers EH. Integrated pest management programs for pear psylla, *Cacopsylla pyricola* (Forster) (Hemiptera: Psyllidae), using kaolin clay and reflective plastic mulch. Journal of Economic Entomology. 2022;**115**:1607-1619. DOI: 10.1093/ee/toac121

[72] Daniel C, Grunder J. Integrated management of European cherry fruit fly *Rhagoletis cerasi* (L.): Situation in Switzerland and Europe. Insects. 2012;**3**:956-988. DOI: 10.3390/insects3040956

[73] Knight AL, Basoalto E, Yee W, Hilton R, Kurtzman CP. Adding yeasts with sugar to increase the number of effective insecticide classes to manage *Drosophila suzukii* (Matsumura) (Diptera: Drosophilidae) in cherry. Pest Management Science. 2016;**72**:1482-1490. DOI: 10.1002/ps.4171

[74] Marshall AT, Melton TD, Bishop G, Clarke AE, Reyes-Corral LA, Catron KA, et al. Cultural control methods improve management of leafhopper vector of X-disease. Crop Protection. 2024;**175**:106445. DOI: 10.1016/J.croppro.2023.106445

[75] Vincent C, Hallman G, Panneton B, Fleurat-Lesand F. Management of agricultural insects with physical control methods. Annual Review of Entomology. 2003;**48**:261-281. DOI: 10.1146/annurevento.091801.112639

[76] Puterka GJ, Glenn DM, Pluta RC. Action of particle films on the biology and behavior of pear psylla (Homoptera: Psyllidae). Journal of Economic Entomology. 2005;**98**:2079-2088. DOI: 10.1093/ee/29.2.329

[77] Unruh TR, Knight AL, Upton J, Glenn DM, Puterka GJ. Particle films for suppression of the codling moth (Lepidoptera: Tortricidae) in apple and pear orchards. Journal of Economic Entomology. 2000;**93**:737-743. DOI: 10.1603/0022-0493-93.3.737

[78] Knight AL, Christianson BA, Unruh TR, Puterka G, Glenn DM. Impacts of seasonal kaolin particle films on apple pest management. Canadian Entomologist. 2001;**133**:413-428. DOI: 10.4039/Ent133413-3

[79] Basoalto E, Miranda M, Knight AL, Fuentes-Contreras E. Landscape analysis of adult codling moth (Lepidoptera: Tortricidae) distribution and dispersal within typical agroecosystems dominated by apple production in central Chile. Environmental Entomology. 2010;**39**:1399-1408. DOI: 10.1603/EN09371

[80] Ioriatti C, Lucchi A. Semiochemical strategies for tortricid control in apple orchards and vineyards in Italy. Journal of Chemical Ecology. 2016;**42**:571-583. DOI: 10.1007/s10886-016-0722-y

[81] Chouinard G, Veilleux J, Pelletier F, Larose M, Philion P, Cormier D. Impact of exclusion netting row covers on arthropod presence and crop damage to 'Honeycrisp' apple trees in North America; a five-year study. Crop Protection. 2017;**98**:248-254. DOI: 10.1016/j.cropro.2017.04.008

[82] Mupambi G, Layne DR, Kalcsits LA, Musacchi S, Serra S, Schmidt T, et al. Use of Protective Netting in Washington State Apple Production [Internet]. 2019. Available from: https://treefruit.wsu.edu/publications/use-of-protective-netting-in-washington-state-apple-production/ [Accessed: February 15, 2025]

[83] Reyes M, Franck P, Olivares J, Margaritopoulos J, Knight A, Sauphanor B. Worldwide variability of insecticide resistance mechanisms in the codling moth, *Cydia pomonella* L. (Lepidoptera: Tortricidae). Bulletin of Entomological Research. 2009;**99**:359-369. DOI: 10.1017/S0007485308006366

[84] Romet L, Severac G, Warlop F. Overview of "ALT'CARPO" concept and its development in France. In: Proceedings of the 14th International Conference in Organic Fruit-Growing (Eco-Fruit); 22-24 February 2010. Germany: University of Hohenheim; 2010. pp. 176-182

[85] Kelderer M, Casera C, Lardscheider E, Rainer A. Controlling codling moth with different netting structures and their influence on crop yield and quality. In: Proceedings of the 14th International Conference in Organic Fruit-Growing (Eco-Fruit); 22-24 February 2010. Germany: University of Hohenheim; 2010. pp. 183-190

[86] Preti M, Basoalto E, Knight AL. Tree architecture impacts the effectiveness of exclusion netting for *Cydia pomonella* mating suppression in pome fruit. Bulletin of Insectology. 2024;**77**(2):193-205

[87] Marshall AT, Beers EH. Exclusion netting affects apple arthropod communities. Biological Control. 2023;**165**:104805. DOI: 10.1016/j.biocontrol.2021.104805

[88] Buck N, Fountain MT, Potts SG, Garratt MPD. Insect excluding mesh enhances spotted-wing drosophila control in tunneled raspberry with limited effects on natural enemy abundance. Agriculture, Ecosystems & Environment. 2024;**359**:108756. DOI: 10.1016/j.agee.2023.108756

[89] Shaw B, Nagy C, Fountain MT. Organic control strategies for use in IPM of invertebrate pests in apple and pear orchards. Insects. 2021;**12**:1106. DOI: 10.3390/insects12121106

[90] Brady M, Granststein D, Kirby E. Survival and growth of organic farms over the long-run. Journal of Agricultural

and Applied Economics Association. 2023;**2**:248-262. DOI: 10.1002/jaa2.53

[91] Orpet RJ, Jones VP, Beers EH, Reganold JP, Goldberger JR, Crowder DW. Perceptions and outcomes of conventional vs. organic apple orchard management. Agriculture, Ecosystems & Environment. 2020;**289**:106723. DOI: 10.1016/j.agee.2019.106723

[92] Jones VP, Horton DR, Mills NJ, Unruh TR, Miliczky E, Shearer PW, et al. Using plant volatile traps to develop phenology models for natural enemies: An example using *Chrysopa nigricornis*. Journal of Biological Control. 2016;**102**:77-84. DOI: 10.1016/j.biocontrol.2014.12.012

[93] Horton DR. Psyllids in natural habitats as alternative resources for key natural enemies of the pear psyllids (Hemiptera: Psylloidea). Insects. 2024;**15**:37. DOI: 10.3390/insects15010037

[94] Gundreddy RR, Alekhya G, Madhuri V, Darjee S, Reddy BT, Gaddam NR. Actuation drones in agriculture: Advancing precision pest management through biocontrol and modern techniques. Journal of Experimental Agriculture International. 2024;**46**:825-835. DOI: 10.9734/jeai/2024/v46i92879

[95] Quarrell SR, Corkrey R, Allen GR. Cherry damage and the spatial distribution of European earwigs, (*Forficula auricularia* L.) in sweet cherry trees. Pest Management Science. 2020;**77**:159-167. DOI: 10.1002/ps.6003

[96] Hanel A, Orpet RJ, Hilton R, Nottingham L, Northfield TD, Schmidt-Jeffris R. Turning a pest into a natural enemy: Removing earwigs from stone fruit and releasing them into pome fruit enhances pest control. Insects. 2023;**14**:906. DOI: 10.3390/insects14120906

[97] Alins G, Lordan J, Rodriguez-Gaol N, Arno J, Penalver-Cruz A. Earwig releases provide accumulative biological control of the woolly apple aphid over the years. Insects. 2023;**14**:890. DOI: 10.3390/insects14110890

[98] Franco JC, Cocco A, Lucchi A, Mendel Z, Suma P, Vacas S, et al. Scientific and technological developments in mating disruption of scale insects. Entomologia Generalis. 2022;**42**:251-273. DOI: 10.1127/entomologia/2021/1220

[99] Maas J, Huang J, Leach H, Wilson JK. Evaluating pheromone dispenser density and longevity for San Jose Scale mating disruption in apple orchards. Crop Protection. 2023;**172**:106337. DOI: 10.1016/j.crop.2023.106337

[100] Guedot C, Millar JG, Horton DR, Landolt PJ. Identification of a sex attractant pheromone for male winterform pear psylla, *Cacopsylla pyricola*. Journal of Chemical Ecology. 2009;**35**:1437-1447. DOI: 10.1007/s10866-009-9725-2

[101] Ganassi S, Germinara GS, Pati S, Civolani S, Cassanelli S, Sabatii MA, et al. Evidence of a female-produced sex pheromone in the european pear psylla, *Cacopsylla pyri*. Bulletin of Insectology. 2018;**71**:57-64. ISSN 1721-8861

[102] Cooper WR, Horton DR, Wildung MR, Jensen AS, Thinakaran J, Rendon D, et al. Host and non-host 'whistle stops' for psyllids: Molecular gut content analysis by high-throughput sequencing reveals landscape-level movements of Psylloidea (Hemiptera). Environmental Entomology. 2019;**48**:554-566. DOI: 10.1093/ee/nvz038

[103] Pajač Beus M, Lemić D, Skendžić S, Čirjak D, Pajač ŽI.

The brown marmorated stink bug (Hemiptera: Pentatomidae) - A major challenge for global plant production. Agriculture. 2024;**14**:1322. DOI: 10.3390/agriculture14081322

[104] Hadden W, Brewster CC, Leskey TC, Bergh JC. *Halyomorpha halys* (Hemiptera: Pentatomidae) trap captures at orchard and nonorchard sites and the influence of uncultivated woody host plants in adjoining woodlots. Journal of Economic Entomology. 2023;**116**:2076-2084. DOI: 10.1093/jee/toad190

[105] Acebes-Doria AL, Morrison WR III, Short BD, Rice KB, Bush HG, Kuhar TP, et al. Monitoring and biosurveillance tools for the brown marmorated stink bug, *Halyomorpha halys* (Stål) (Hemiptera: Pentatomidae). Insects. 2018;**9**:82. DOI: 10.3390/insects9030082

[106] Short BD, Khrimian A, Leskey TC. Pheromone-based decision support tools for management of *Halyomorpha halys* in apple orchards: Development of a trap-based treatment threshold. Journal of Pest Science. 2016;**90**:1191-1204. DOI: 10.1007/s10340-016-0812-1

[107] Forresi C, Gallinucci E, Golfarelli M, Maistrello L, Preti M, Vaccari G. A data platform for real-time monitoring and analysis of the brown marmorated stink bug in northern Italy. Ecological Informatics. 2024;**82**:102713. DOI: 10.1016/j.ecoinf.2024.102713

[108] Rice KB, Cullum JP, Wiman NG, Hilton R, Leskey TC. *Halyomorpha halys* (Hemiptera: Pentatomidae) response to pyramid traps baited with attractive light and pheromonal stimuli. Florida Entomologist. 2017a;**100**:449-453. DOI: 10.1653/024.100.0207

[109] Rondoni G, Chierici E, Marchetti E, Nasi S, Ferrari R, Conti E. Improved captures of the invasive brown marmorated stink bug, *Halyomorpha halys*, using a novel multimodal trap. Insects. 2022;**13**:527. DOI: 10.3390/insects13060527

[110] Zapponi L, Nieri R, Zaffaroni-Caorsi V, Pugno NM, Mazzoni V. Vibrational calling signals improve the efficacy of pheromone traps to capture the brown marmorated stink bug. Journal of Pest Science. 2023;**96**:587-597. DOI: 10.1007/s10340-022-01533-0

[111] Kargar A, Zorbas D, Tedesco S, Gaffney M, O'Flynn B. Detecting *Halyomorpha halys* using a low-power edge-based monitoring system. Computers and Electronics in Agriculture. 2024;**221**:108935. DOI: 10.1016/j.compag.2024.108935

[112] Giannetti D, Patelli N, Palazzetti L, Betti Sorbelli F, Pinotti CM, Maistrello L. First use of unmanned aerial vehicles to monitor *Halyomorpha halys* and recognize it using artificial intelligence. Pest Management Science. 2024;**80**:4074-4084. DOI: 10.1002/ps.8115

[113] Kuhar TP, Kamminga K. Review of the chemical control research on *Halyomorpha halys* in the USA. Journal of Pest Science. 2017;**90**:1021-1031. DOI: 10.1007/s10340-017-0859-7

[114] Akotsen-Mensah C, Blaauw B, Short B, Leskey TC, Bergh JC, Polk D, et al. Using IPM-CPR as a management program for apple orchards. Journal of Economic Entomology. 2020;**113**:1894-1902. DOI: 10.1093/jee/toaa087

[115] Fornasiero D, Scaccini D, Lombardo V, Galli G, Pozzebon A. Effect of exclusion net timing of deployment and color on *Halyomorpha halys* (Hemiptera: Pentatomidae) infestation in pear and apple orchards. Crop Protection. 2023;**172**:106331. DOI: 10.1016/j.cropro.2023.106331

[116] Chierici E, Marchetti E, Poccia A, Russo A, Giannuzzi VA, Governatori L, et al. Laboratory and field efficacy of natural products against the invasive pest *Halyomorpha halys* and side effects on the biocontrol agent *Trissolcus japonicus*. Scientific Reports. 2025;**15**:4622. DOI: 10.1038/s41598-025-87325-9

[117] Kuhar TP, Short BD, Krawczyk GR, Leskey TC. Deltamethrin-incorporated nets as an integrated pest management tool for the invasive *Halyomorpha halys* (Hemiptera: Pentatomidae). Journal of Economic Entomology. 2017;**110**:543-545. DOI: 10.1093/jee/tow321

[118] Masetti A, Morelli A, Fagioli L, Pradolesi G, Nicoli R, Scagliarini O, et al. Evaluation of an attract-and-kill strategy using long-lasting insecticide nets for the management of the brown marmorated stink bug in northern Italy. Insects. 2024;**15**:577. DOI: 10.3390/insects15080577

[119] Carnio V, Favaro R, Preti M, Angeli S. Impact of aggregation pheromone traps on spatial distribution of *Halyomorpha halys* damage in apple orchards. Insects. 2024;**15**(10):791. DOI: 10.3390/insects15100791

[120] Nielsen AL, Dively G, Pote JM, Zinati G, Mathews C. Identifying a potential trap crop for a novel insect pest, *Halyomorpha halys* (Hemiptera: Pentatomidae), in organic farms. Environmental Entomology. 2016;**45**:472-478. DOI: 10.1093/ee/nvw006

[121] Zhu G, Illan JG, Hajek AE, Nielsen AL, Leskey TC, Walgenbach JF, et al. Assessing geographic dimensions of biological control for *Halyomorpha halys* in United States. Entomologia Generalis. 2024;**44**:895-904. DOI: 10.1127/entomologia/2024/2528

[122] Zapponi L, Tortorici F, Anfora G, Bardella S, Bariselli M, Benvenuto L, ..., Sabbatini-Peverieri G. Assessing the distribution of exotic egg parasitoids of *Halyomorpha halys* in Europe with a large-scale monitoring program. Insects. 2021; 12: 316. DOI: 10.3390/insects12040316

[123] Cloonan KR, Abraham J, Angeli S, Syed Z, Rodriguez-Saona C. Advances in the chemical ecology of the spotted wing Drosophila (*Drosophila suzukii*) and its applications. Journal of Chemical Ecology. 2018;**44**:922-939. DOI: 10.1007/s10886-018-1000-y

[124] McIntosh H. *Drosophila suzukii* development and attraction [thesis]. Department of Environmental Science and the Robert D. Clark Honors College of the University of Oregon; 2016. p. 36. Available from: https://hdl.handle.net/1794/20329 [Accessed: February 15,2025]

[125] Kirkpatrick DM, Gut LJ, Miller JR. Development of a novel dry, sticky trap design incorporating visual cues for *Drosophila suzukii* (Diptera: Drosophilidae). Journal of Economic Entomology. 2018;**111**:1775-1779. DOI: 10.1093/jee/toy097

[126] Hong Z, Michel AP, Long EY. Optimizing a rapid LAMP assay for discrimination of *Drosophila suzukii* (Diptera: Drosophilidae) from common drosophilids captured in monitoring traps from the Midwest, United States. Journal of Economic Entomology. 2023;**116**:1391-1397. DOI: 10.1093/jee/toad105

[127] Kalfas I, De Ketelaere B, Belien T, Saeys W. Optical identification of fruit fly species based on their wingbeats using convolutional neural networks. Frontiers in Plant Science. 2022;**13**:812506. DOI: 10.3389/fpls.2022.812506

[128] Bischel DM. Classification of the Sex of *Drosophila suzukii* with Pre-Trained

Networks. Dissertation UC Santa Cruz [Internet]; 2023. Available from: https://escholarship.org/uc/item/1j1173ms [Accessed: January 20, 2025]

[129] Leach H, Moses J, Hanson E, Fanning P, Issacs R. Rapid harvest schedules and fruit removal as non-chemical approaches for managing spotted wing Drosophila. Journal of Pest Science. 2017;**91**:219-226. DOI: 10.1007/s10340-017-0873-9

[130] Kuesel R, Gonthier D. Fine-mesh exclusion netting for control of *Drosophila suzukii*. In: Garcia FRM, editor. *Drosophila suzukii* Management. Cham: Springer; 2020. pp. 217-239. DOI: 10.1007/978-3-030-62692-1_11

[131] Lee JC, Wang X, Daane KM, Hoelmer KA, Issacs R, Sial AA, et al. Biological control of spotted-wing Drosophila (Diptera: Drosophilidae) – Current and pending tactics. Journal of Integrated Pest Management. 2019;**10**:1-9. DOI: 10.1093/jipm/pmz012

[132] Wang X, Daane KM, Hoelmer KA, Lee JC. Biological control of spotted-wing Drosophila: An update on promising agents. In: Garcia FRM, editor. *Drosophila suzukii* Management. Cham: Springer; 2020. pp. 143-167. DOI: 10.1007/978-3-030-62692-1_8

[133] Andika IP, Vandervoort C, Wise JC. Curative activity of insecticides used to control spotted-wing Drosophila in tart cherry productions. Journal of Economic Entomology. 2020;**113**:2372-2379. DOI: 10.1093/jee/toaa161

[134] Noble R, Shaw B, Walker A, Whitfield EC, Deakin G, Harris A, et al. Control of spotted wing drosophila (*Drosophila suzukii*) in sweet cherry and raspberry using bait sprays. Journal of Pest Science. 2023;**96**:623-633. DOI: 10.1007/s10340-022-01566-5

[135] Bjeljac M, Spitaler U, Mori N, Fusillo M, Bombardini E, Preti M, et al. Canopy strip applications of *Hanseniaspora uvarum* combined with spinosad reduce insecticide use without compromising *Drosophila suzukii* control in cherry. Crop Protection. 2024;**186**:106868. DOI: 10.1016/j.cropro.2024.106868

[136] Rice KB, Short BD, Leskey TC. Development of an attract-and-kill strategy for *Drosphila suzukii* (Diptera: Drosophilidae): Evaluation of attracticidal spheres under laboratory ad field conditions. Journal of Economic Entomology. 2017b;**110**:535-542. DOI: 10.1093/jee/tow319

[137] Yousef M, Aranda-Valera E, Quesada-Moraga E. Lure-and-infect and lure-and-kill devices based on *Metarhizium brunneum* for spotted wing Drosophila control. Journal of Pest Science. 2017;**91**:227-235. DOI: 10.1007/S10340-017-0874-8

[138] Nixon LJ, Cloonan K, Rugh A, Jones S, Evans B, Rice K, et al. Factors affecting the efficacy of attracticidal spheres for management of *Drosophila suzukii* (Diptera: Drosophilidae). Journal of Applied Entomology. 2021;**146**:243-251. DOI: 10.1111/jen.12961

[139] Schetelig MF, Schwirz J, Yan Y. A transgenic female killing system for the genetic control of *Drosophila suzukii*. Scientific Reports. 2021;**11**:12938. DOI: 10.1038/s41598-021-91938-1

[140] Homem RA, Mateos-Fierro Z, Jones R, Gilbert D, Mckemey AR, Slade G, et al. Field suppression of spotted wing Drosophila (SWD) (*Drosphila suzukii* Matsumura) using the sterile insect technique (SIT). Insects. 2022;**13**:328. DOI: 10.3390/insects13040328

Chapter 5

Classification of Insect Parasitic Nematodes and Potential of Entomopathogenic Nematodes as Biocontrol Agents against Insect Pests in Agriculture

Heena Khan, Nitesh Kumar and Ashok Kumar Chaubey

Abstract

Agricultural crops are subject to serious, significant, and widespread damage from attacks by a wide variety of harmful destructive pest species. Insect pests are the primary biotic stressors for crops and significantly reduce agricultural production. The herbivorous insects are responsible for the destruction of approximately one-fifth of global crop production each year. For controlling these insect pests, insecticides play a vital role in Integrated Pest Management (IPM), through which insects can be destroyed or the specific developmental stages of insects can be inhibited but with time, these insects develop resistance against synthetic chemical insecticides which is a very common problem in effective control of insect pests. Over the past few decades, interest in the use of nematodes as biological control of insect pests has increased dramatically throughout the globe. Thus, the classification of insect parasitic nematodes and the use of entomopathogenic nematodes as potential biocontrol agent against harmful insect pests are discussed in this chapter.

Keywords: insect parasitic nematodes, entomopathogenic nematodes, integrated pest management, *Steinernema*, *Heterorhabditis*, biocontrol

1. Introduction

Agricultural crops have been facing distressing and considerable damage due to attack of a wide variety of destructive pest species. More than 10,000 insect species, 30,000 weed species, 100,000 disease-causing organisms (fungi, viruses, bacteria, and other microbes), and ~ 1000 nematode species are known to cause severe damage to food plants around the world [1]. However, from all the identified pest species, less than 10% of species are considered major pests [2]. In relation to the insects, some species are beneficial but many of them are considered major pests which are harmful to the various crops. Insects infest on different parts of agricultural crops and cause significant yield losses.

1.1 Insects as the primary biotic stressors for the crops and cause of economic loss in agriculture

Insect pests are the primary biotic stressors for the crops and significantly reduce the agricultural production. Based on the feeding habits of insects, they can be monophagous or polyphagous. The monophagous insect pest feeds on a single crop or grain while the polyphagous insect feeds on a variety of plant tissues and grains. The direct economic losses in the agricultural sector are associated with the consumption of dry matter, while indirect losses are caused by the contamination of damaged kernels and processed products. It is estimated that herbivorous insects are responsible for the destruction of approximately one-fifth of global crop production each year. There has been a dramatic rise in the economic losses caused by insect pests in various agricultural crops, where the cotton crop suffers maximum losses (30%), followed by rice (25%), sugarcane (20%), rapeseed-mustard (20%), maize (18%), groundnut and pulses (15%), other oilseeds (12%), coarse cereals (8%), and wheat (5%) [2].

1.2 Direct and indirect damages caused by insect pests in agricultural crops

Insects are responsible for two main types of damage to developing crops and cause significant loss in agricultural production. The first is direct damage to the plant caused by the feeding insect, which burrows in stems, fruit, or roots or consumes leaves. Among the orthopterans, homopterans, heteropterans, coleopterans, lepidopterans, and dipterans, there are hundreds of species of this kind of insect pest, both as adults and as larvae. The second type is indirect damage, where an insect pest itself causes little or no injury to the crop but carries diseases and infects a crop with a bacterial, viral, or fungal infection. For instance, aphids spread the viral diseases of potatoes and sugar beets from one plant to another.

1.3 Insects as destructive pests of crops, stored grains, and their products

The storage insects can be classified as primary and secondary pests based on their feeding habits. Primary insect pests are more destructive because they have the ability to attack intact plant tissues and crop derivatives. These insect pests typically perforate the plant and grain to feed and reproduce. They mainly feed on plant tissues such as tissues of roots, shoots, fruits, and kernels and use these tissues as the ideal site for oviposition and subsequent growth and development of the larvae. The primary insect pests are found within the order Orthoptera, Lepidoptera, Coleoptera, and Hymenoptera. On the other hand, secondary insect pests are opportunistic in nature. They feed on plant tissues that are damaged or destroyed by primary insects, or by processed products like flours or grits or end products. Primary and secondary insects both use plant tissues as an ideal habitat for their life and reproduction. Globally, the crop-damaging insect group is dominated by the group of moths, which includes cutworm, armyworm, earworm, borer, and grain moth. The beetles (rootworm, wireworm, grub, grain borer, and weevil) are the second most destructive insect pests to crops. The group of insects that act as vectors for disease agents includes leafhoppers and aphids.

1.4 Insecticides in pest management

Insecticides play a vital role in controlling the insect pests under Integrated Pest Management (IPM), through which the insects can be destroyed or the specific developmental stages of insects can be inhibited. The first large-scale application of pesticide (Paris green) in agriculture occurred in the 1860s as a result of massive Colorado potato beetle (a major pest of potato crop) infestations. Later, in the 1940s, other insecticides containing highly poisonous chemicals such as lead arsenate, calcium arsenate, and concentrated nicotine were also used in large quantities for controlling insect pests. But with time, the insects develop resistance against these synthetic chemical insecticides which is a very common problem in effective control of insect pests. As a result of which, the use of chemical pesticides has decreased and the biological methods of pest control have become more popular. Thus, over the past few decades, the interest in the use of nematodes as biocontrol agents to reduce the population of insect pests has increased dramatically throughout the globe.

2. Insect parasitic nematodes

Nematodes have established a trophic relationship with insects. The association of nematodes with insects opens the door for these nematodes to migrate to new habitats as their insect hosts broaden their range, either through climate change or human activity [3]. The life-history strategies of nematode species that attack insects are varied. Some live inside the insect host like parasites but do not kill their host. Others have symbiotic bacteria that help them to kill the host quickly. Many parasitic nematodes undergo a "pre-parasitic" or "non-parasitic" phase where they do not need a host. In some cases, this period is spent within the eggshell where the enclosed juvenile waits to be eaten by the host. However, in many cases, juveniles hatch out from eggshells as free-living forms and utilize all the necessary natural resources which help them to sustain growth and development and also to establish the parasitic relationship with the host. They show different types of relationships with insects such as symbiosis, phoresis, obligate, and facultative parasitism [4]. Nematodes possess some capacity to find or attack the hosts. Many of them can enter into the insect's body through their natural openings (such as mouth, spiracles, and anus) or wounds and often penetrate through rupturing the body cavity (haemocoel) of insects. Other nematodes can enter the insect gut passively when the nematodes or their eggs are ingested with food by the insect. Some nematodes bear teeth-like projections in their mouth and use them to penetrate the host cuticle and actively enter inside the insect host. However, the nematodes lacking teeth can also enter through the cuticle because they are equipped with some glands filled with tissue-damaging enzymes. Once they find the insect host, they can cause sterility, delayed development, reduced fecundity, aberrant behavior, or even death of the insect host by affecting their host's morphology and physiology. Some of them find and infect adult female insects and reach their ovary or eggs to colonize there. They do not kill their female insect host but take advantage of oviposition to spread themselves to other places where healthy immature insects would present. During the egg laying, the parasitized female deposits the nematode's infective juveniles (IJs) instead of host eggs or lays the eggs along with IJs. Whereas, some nematodes (*Steinernema* and *Heterorhabditis*) are the natural enemies of insects, who actively seek out and locate insect hosts using volatile cues such as fecal matter of host, gradients of bacteria, carbon dioxide, and chemical signals from damaged plant roots. They enter the insect host and kill their host by producing toxins and suppressing their immune system and are known as entomopathogenic nematodes (EPNs). They simultaneously evade the host immune system and reduce the insect population by parasitism.

Insect parasitic nematodes can be classified on the basis of the following different criteria-

i. Classification based on Lifestyle or Lifecycle;

ii. Classification based on Disease-causing Ability; and

iii. Classification based on Parasitic Strategies

2.1 Classification based on lifestyle or lifecycle

The parasitic nematodes have a wide variety of lifestyles or lifecycles. They are able to infect almost all species and families of organisms and are widely distributed in marine, land, and freshwater environments. All the parasitic nematodes depend on the host for their survival but due to their different lifestyle or lifecycle, the dependency of nematodes on host can be varied. Depending on the host species, they may be highly specific to the host species or non-specific to the host species, and may or may not have free-living phases or intermediate hosts [5]. Based on the lifestyle or lifecycle, the insect-parasitic nematodes can be classified into two categories-.

2.1.1 Facultative parasitic nematodes

These parasitic nematodes do not depend on insect host for completing their lifecycle but once they come in contact with insect host, they become parasites. As they are independent, they have their own metabolic pathway and simple lifecycle. They are capable of reproducing outside the host. Several nematode species of the order Rhabditida and Tylenchida are facultative parasites of insects. The nematode *Beddingia siricidicola* (order Tylenchida) complicates its lifecycle in both ways, i.e., parasitic as well as free-living generations. They infect woodwasp (*Sirex noctilio*), but in the absence of the woodwasp, they continue their lifecycle as free-livings. They feed on the fungus *Amylostereum chailletti,* which occurs on the tree trunk and this is the only fungus consumed by *Beddingia siricidicola* nematode [6]. In the family Rhabditidae, most of the members primarily inhabit decaying organic matter in soil. One of the members of this family is *Rhabditis insectivora*, which is a facultative parasite that occurs as juveniles and adults in the haemocoel of the antelope beetle *Dorcus parallelipipedus*. *Panagrellus ceylonensis* is a bacterial-feeder free-living nematode but this nematode can enter the insect's body, establish a colony there, and live inside the gut of insects as a facultative parasite [7]. They can reproduce inside the insect host without causing any harm to the insect host (e.g., about 48 nematodes including adults and juveniles were recovered from the gut of a live and healthy butterfly *M. leda*) [7].

2.1.2 Obligate parasitic nematodes

Such parasitic nematodes are completely dependent on the host in order to complete their lifecycle. They do not have their own metabolic pathway and lead only parasitic life. As they are dependent on host, they have a complex lifecycle in which they are incapable of reproducing outside the host, due to which they require insect host for completing their lifecycle. *Paraiotonchium autumnale* (Tylenchida) is an obligatory insect parasitic nematode which infects and sterilizes their insect host rather than killing immediately, whereas *Steinernema* spp. and *Heterorhabditis* spp. (Rhabditida) are the lethal obligatory parasites of insects.

2.2 Classification based on disease-causing ability

On the basis of disease-causing ability, the insect parasitic nematodes can be classified as mentioned below-.

2.2.1 Pathogenic or lethal parasitic nematodes

The parasitic nematodes not only deprive nourishment from the insect host but also cause disease to the host and are capable of killing them. Most of the species of *Steinernema* and *Heterorhabditis* are pathogenic or lethal to various insect hosts.

2.2.2 Non-pathogenic or non-lethal parasitic nematodes

The nematodes of this category do not cause disease to their host and develop within the living insect only to deprive nourishment from them. They are not lethal to their host because they do not have ability to kill their host. *Panagrellus ceylonensis* is a non-pathogenic or non-lethal insect parasitic nematode which does not cause disease to the insect host but can live inside the insect host to deprive nourishment from them.

2.2.3 Opportunistic parasitic nematodes

Normally, these parasitic nematodes do not cause disease to their insect host but under certain conditions such as impaired host immunity, they become pathogenic. *Distolabrellus veechi* nematodes are non-pathogenic nematodes, they usually do not cause disease to their insect host but can parasitize and kill their insect host in some cases.

2.3 Classification based on parasitic strategies

Based on the parasitic strategies, insect-parasitic nematodes can be classified into the following four categories-.

2.3.1 Parasitic castrators

The parasitic nematodes use parasitic castration strategy to destroy their host's ability of reproduction partly or completely. An insect-parasitic nematode *Howardula husseyi* (Tylenchida: Allantonematidae) is a naturally occurring nematode of phorid fly (*Megaselia halterata*) which enters the fly larva and reproduces inside the host. Then, they develop inside the fly and start consuming the fly's reproductive system. An infected female fly produces only a few offspring if available and when the infected female fly attempts to lay eggs, these nematodes are released back into the compost and casing. These released parasitic nematodes are again ready to infect other phorid larvae [8]. Other than this, an insect-parasitic nematode *Sphaerularia bombi* also alters the behavior of queens in different species of bumblebee (*Bombus* spp.) by infecting them and destroying their reproductive system through a parasitic castration strategy.

2.3.2 Directly transmitted parasitic nematodes

These parasitic nematodes use host-finding strategies and reach the insect host on their own without using any vector for transmission. They infect only a single insect host in their lifetime. Usually, they cause mild to moderate pathology, but there are a few directly transmitted parasites that can be quite virulent. Insect-parasitic nematodes of the order Rhabditida and Tylenchida are examples of directly transmitted parasites. They enter inside the body of insect host through their natural openings (*Steinernema* spp. and *Heterorhabditis* spp.) or pierce the body wall of an insect host and gain direct access to the haemocoel (*Hexatylina* spp.) which is an extraordinary adaptation in parasitism.

2.3.3 Trophically transmitted parasitic nematodes

In this type of parasitic strategy, the parasitic nematodes passively enter the insect gut when these nematodes are eaten by a host. This is considered to be one of the most significant adaptive peaks for parasitism [9–12]. *Streptopharagus pigmentatus* is a spiruroid nematode which uses several species of dung beetle as their intermediate host and can infect a variety of Old World Monkeys, Macaca, Papio, etc.

2.3.4 Vector transmitted parasitic nematodes

These parasitic nematodes are dependent on the intermediate host (vector) to reach their definitive host. They always require two hosts to complete their lifecycle. The nematode *Bursaphelenchus cocophilus* is a primary agent that causes red-ring disease of coconuts and palms but it can be vectored by palm weevil *Rhynchophorus palmarum* (L.) [13, 14].

3. Insect parasitic nematodes and their association with bacterial species

Insect parasitic nematodes generally occur in the soil and establish close association with some bacterial species. Nematode species that attack insects have diverse life-history strategies where some of them feed on the insect host as parasites, but do not kill their host and others are associated with symbiotic bacteria that help them to cause rapid insect death. The nematodes belonging to the order Rhabditida have great ecological diversity and they are closely related to insects, which makes them important for nematological research.

Nematodes belong to the families Steinernematidae and Heterorhabditidae are unique insect parasites because they are symbiotically associated with entomopathogenic bacteria, due to which they are considered potential biocontrol agents to control insect pests. Furthermore, other soil-borne insect parasitic nematodes such as *Oscheius* and *Acrobeloides* species are also found in close associations with several bacteria. Although *Oscheius* species are facultative parasites, they can be lethal to some insect pests. They established association with entomopathogenic bacteria belonging to the genera *Serratia* [15–17]. Whereas *Acrobeloides* species showed association with different bacterial genera including *Bacillus, Alcaligenes, Enterobacter, Klebsiella,* and *Pseudomonas* [17].

Moreover, several other bacterivorous nematodes that are associated with insects belong to the families *Panagrolaimidae, Mesorhabditidae,* and *Cephalobidae*.

Nematodes of the *Panagrellus* genus Thorne, 1938 (family Panagrolaimidae), are bacterial-feeder and free-living in nature. Most of the *Panagrellus* species are discovered from bacterial-rich environments, but these nematodes are also associated with insects as facultative parasites. *Distolabrellus veechi* is also a bacterivorous free-living nematode which also could parasitize insects [18] and belongs to the family Mesorhabditidae. Other rhabditids have been reported to possess this characteristic, even though they are typically found on (as ectoparasites) or in insects (as endoparasite), and can develop on an artificial medium without the requirement of a host [19–21]. A list of insect parasitic nematodes and their associated bacterial species is mentioned in **Table 1**.

Nematode species	Bacterial species	Reference
Steinernema	***Xenorhabdus***	
S. abbasi	*X. indica*	Tsai et al. [22]
S. aciari	*X. ishibashii*	Kuwata et al. [23]
S. affine	*X. bovienii*	Akhurst and Boemare [24]
S. anatoliense	*X. nematophila*	Stock et al. [25]
S. apuliae	*X. kozodoii*	Tailliez et al. [26]
S. arenarium	*X. kozodoii*	Tailliez et al. [26]
S. australe	*X. magdalenensis*	Tailliez et al. [27]
S. bicornutum	*X. budapestensis*	Lengyel et al. [28]
S. beitlechemi	*X. khoisanae*	Cimen et al. [29]
S. biddulphi	*X. indica*	Cimen et al. [30]
S. boemarei	*X. kozodoii*	Tailliez et al. [26]
S. ceratophorum	*X. budapestensis*	Yang et al. [31]
S. costaricense	*X. szentirmaii*	Lengyel et al. [28]
S. cubanum	*X. poinarii*	Fischer-Le Saux et al. [32]
	X. romanii	Tailliez et al. [26]
S. carpocapsae	*X. nematophila*	Akhurst and Boemare [24]; Martens and Goodrich-Blair [33]; Stock et al. [25]
S. citrae	*X. bovienii*	Stokwe et al. [34]
S. diaprepesi	*X. doucetiae*	Tailliez et al. [26]
S. eapokense	*X. eapokensis*	Kämpfer et al. [35]
S. fabii	*X. khoisanae*	Abate et al. [36]
S. feltiae	*X. bovienii*	Akhurst and Boemare [24]; Ehlers et al. [37]; Stock et al. [25]
S. glaseri	*X. poinarii*	Akhurst [38]; Akhurst and Boemare [24]
S. hermaphroditum	*X. griffniae*	Tailliez et al. [26]
S. ichnusae	*X. bovienii*	Tarasco et al. [39]
S. intermedium	*X. bovienii*	Akhurst [38]; Akhurst and Boemare [24]
S. jefreyense	*X. khoisanae*	Dreyer [40]

Nematode species	Bacterial species	Reference
S. khoisanae	*X. khoisanae*	Ferreira et al. [41]
S. karii	*X. hominickii*	Tailliez et al. [26]
S. kushidai	*X. japonicus*	Nishimura et al. [42]
S. kraussei	*Xenorhabdus sp.*	Akhurst [43]
	X. bovienii	Akhurst and Boemare [24]
S. khoisanae	*X. khoisanae*	Ferreira et al. [41]
S. khuongi	*X. poinarii*	Baniya and DiGennaro [44]
S. longicaudum	*X. ehlersii*	Lengyel et al. [28]
	X. beddingii	Akhurst and Boemare [24]
S. litorale	*X. bovienii*	Özdemir et al. [45]
S. litchii	*X. grifniae*	Dreyer [40]
S. monticolum	*Xenorhabdus sp.*	Kang et al. [46]
	X. hominickii	Tailliez et al. [26]
S. poinari	*X. bovienii*	Sajnaga et al. [47]
S. puertoricense	*X. romanii*	Tailliez et al. [26]
S. rarum	*X. szentirmaii*	Lengyel et al. [28]
S. riobrave	*X. cabanillasii*	Tailliez et al. [26]
S. sangi	*X. thuongxuanensis*	Kämpfer et al. [35]
	X. vietnamensis,	Kämpfer et al. [35]; Tailliez et al. [48]
S. scapterisci	*X. innexi*	Lengyel et al. [28]; Kim et al. [49]
S. scarabaei	*X. koppenhoeferi*	Tailliez et al. [26]
S. serratum	*X. ehlersii*	Lengyel et al. [28]
S. siamkayai	*X. stockiae*	Tailliez et al. [26]; Ardpairin et al. [50]
S. sacchari	*X. khoisanae*	Dreyer [40]
S. surkhetense	*X. stockiae*	Bhat et al. [51]
S. silvaticum	*X. bovienii*	Akhurst and Boemare [24]
Steinernema spp.	*X. miraniensis,* *X. mauleonii*	Tailliez et al. [26]
S. tbilisiensis	*X. bovienii*	Gorgadze et al. [52]
S. unicornum	*X. lircayensis*	Castaneda-Alvarez et al. [53]
S. vulcanicum	*X. kozodoii*	Clausi et al. [54]
S. yirgalemense	*X. indica*	Tamiru et al. [55]
Heterorhabditis	***Photorhabdus***	
H. atacamensis	*P. khanii subsp. guanajuatensis*	Machado et al. [56]

Nematode species	Bacterial species	Reference
H. bacteriophora	*P. luminescens*	Boemare et al. [57]; Stock et al. [25]
	P. luminescens subsp. laumondii	Tailliez et al. [48]
	P. luminescens subsp. kayaii	
	P. khanii	Tailliez et al. [48]; Machado et al. [58]
	P. luminescens subsp. thracensis	Machado et al. [58]
	P. luminescens subsp. luminescens	
	P. kayaii	
	P. caribbeanensis	
	P. laumondii	
	P. stackebrandtii	
	P. thracensis	
	P. aegyptia	Machado et al. [59]
	Morganella morganii subsp. morganii	Loulou et al. [17]
H. baujardi	*P. namnaonensis*	Glaeser et al. [60]
	P. luminescens	
H. beicherriana	*P. bodei*	Machado et al. [58]
H. casmirica	*Photorhabdus laumondii subsp. clarkei*	Bhat et al. [61]
H. downesi	*P. namnaonensis*	Glaeser et al. [60]
	P. cinerea	Machado et al. [58]
	P. temperata	
H. floridensis	*P. luminescens subsp. luminescens*	Blackburn et al. [62]
H. georgiana	*P. luminescens subsp. akhurstii*	Machado et al. [58]
	P. stackebrandtii	
	P. kleinii	Plichta et al. [63]
H. gerrardi	*P. asymbiotica*	Akhurst et al. [64]
	P. australis subsp. australis	Plichta et al. [63]
	P. luminescens subsp. thailandensis	Machado et al. [59]
	P. australis	Machado et al. [58]
H. indica	*P. akhurstii*	Machado et al. [58]

Nematode species	Bacterial species	Reference
H. indicus	*P. luminescens subsp. akhurstii*	Machado et al. [59]
	P. asymbiotica	
	P. aegyptia	
H. marelatus	*P. tasmanensis*	Machado et al. [58]
H. maxicana	*P. luminescens subsp. mexicana*	Machado et al. [56]
H. megidis	*P. temperate*	Toth and Lakatos [65]
	P. temperata	Fischer-Le Saux et al. [32]
	P. cinerea	Machado et al. [58]
H. noenieputensis	*P. luminescens subsp. noenieputensis*	Ferreira et al. [41]
	P. noenieputensis	Machado et al. [58]
H. safricana	*P. luminescens subsp. laumondii*	Geldenhuys et al. [66]
H. sonorensis	*P. luminescens subsp. sonorensis*	Orozco et al. [67]
H. zealandica	*P. heterorhabditis*	Ferreira et al. [68]
	P. tasmaniensis	Machado et al. [58]
Heterorhabditis sp.	*P. hainanensis*	Tailliez et al. [48]; Machado et al. [58]
	P. hindustanensis	Machado et al. [69]
Oscheius	**Bacteria**	
O. chongmingensis	*Serratia nematodiphila*	Zhang et al. [70]; Zhan et al. [71]
	Ochrobactrum tritici	Fu & Liu [72]
	Bacillus cereus	
O. carolinensis	*Serratia marcescens*	Torres-Barragan et al. [73]
	Enterococcus mundti	
	Achromobacter xylosoxidans	
	Providencia rettgeri	
O. safricana	*Serratia marcescens*	Torres-Barragan et al. [73]
O. basothovii	*Serratia sp.*	Lephoto et al. [16]
O. microvilli	*Serratia sp.*	Zhou et al. [74]
Rhabditis	**Bacteria**	
Rhabditis blumi	*Alcaligenes faecalis*	Park et al. [75]
	Flavobacterium sp.	
	Providencia vermicola	

Table 1.
Insect parasitic nematodes and their associated bacterial species.

4. Entomopathogenic nematodes: Lethal obligatory parasites with bacterial symbionts

Entomopathogenic nematodes (EPNs) are the soil-inhabiting, lethal obligate parasites of a wide range of insects worldwide which kill their host rapidly. These nematodes are symbiotically associated with entomopathogenic bacteria belonging to the family Enterobacteriacae, which makes them lethal obligatory insect parasites that affect a variety of insects including their larval forms and kill them within a short period of time. The first entomopathogenic nematode was described in 1923 [76, 77], since then more than 130 EPN species have been described so far from different geographical regions. EPNs have been described from 23 families of nematodes [78] but the EPNs, belonging to the families Steinernematidae and Heterorhabditidae, are the most promising and highly effective biological control agents against many insect pests [79]. The two major genera *Steinernema* and *Heterorhabditis* of these families are symbiotically associated with the bacteria of the genus *Xenorhabdus* and *Photorhabdus*, respectively. These symbiotic bacteria produce secondary metabolites that cause cytotoxic activity and suppress the immune system of insect resulting in septicaemia and death of host. The secondary metabolites produced by symbiotic bacteria also prevent the development of other bacteria, fungi, and protists in the abundant haemolymph of a dead insect (cadaver), making it an ideal environment for nematodes to multiply [80, 81]. If the color of infected cadaver becomes brown or tan, it is suspected that the insect was killed by Steinernematid nematodes, while in case of Heterorhabditid nematode, the color of cadaver turned into brick red [79]. This color difference in body of insect is caused by the pigment produced by the luminescence bacteria *Photorhabdus* [79, 82]. Recently, a bacterial strain *Morganella morganii* subsp. *morganii* was isolated from *Heterorhabditis bacteriophora* in Germany [17].

Other than *Steinernema* and *Heterorhabditis* species, there are some other nematode species that have been described as EPNs including *Oscheius chongmingensis* (formerly known as *Heterorhabditoides chongmingensis*) [83], *Oscheius carolinensis* [84], *Oscheius gingeri* [85], *Oscheius onirici* [86], and *Caenorhabditis briggsae* [15]. However, Zhang et al. [87] differentiate scavengers and EPNs in their study and concluded that *O. chongmingensis* behaves more like a scavenger than an EPN and is considered a free-living bacterivorous nematode.

5. Lifestyle and survival mechanism of EPNs and free-living bacterivorous nematodes

The lifestyle of EPN and free-living bacterivorous nematodes is different and can be distinguished. The IJs of EPNs are associated with entomopathogenic bacterial symbionts that are passed down through many generations in which the IJs of EPNs are re-associated with their specific symbionts during the development of IJ stage [88]. The IJs either infect the healthy insects and act as true entomopathogens [89] or colonize already-dead insect and act as facultative scavenger [90, 91]. The ability to locate and attack hosts is one of the distinguishing features of EPNs. However, in some cases, EPNs utilize the scavenging tactics as a method of survival [90], where the IJs of EPNs are more attracted towards already-dead or infected hosts than towards live or healthy hosts [92]. After invading the host, they release their bacterial symbiont inside the insect host and utilize them to kill their host. Then they develop into adults inside the host and reproduce. EPNs require insect host in order to complete their lifecycle because they are incapable of reproducing outside the host.

In contrast, free-living bacterivorous nematodes act as scavengers and are mostly attracted towards the dead infected insect and least attracted towards healthy, uninfected insect host (**Figure 1**). They do not depend on the insect host for completion of their lifecycle because they are capable of reproducing outside the host. As they

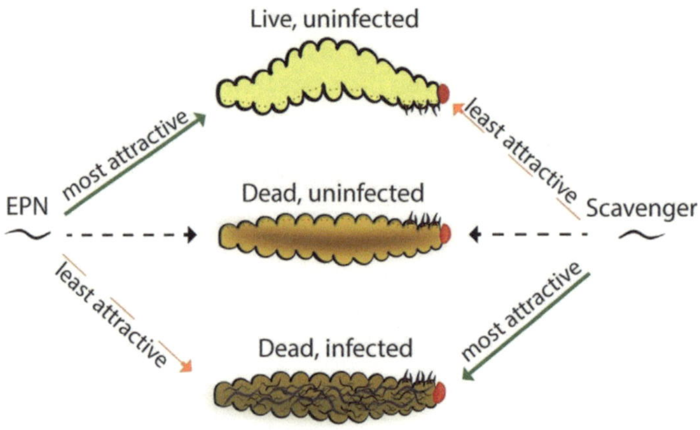

Figure 1.
The survival strategies of EPNs and Scavenger nematodes [87].

are independent, they have their own metabolic pathway and simple lifecycle. Also, they do not possess symbiotically associated pathogenic bacteria, but once they come in contact with insect host, they become parasitic and can act as facultative parasites. Moreover, bacterivorous nematodes play a vital role in many ecological processes, such as nutrient cycling and decomposition processes, thereby indirectly affecting primary production in natural ecosystems. Organic matter provides energy source and enhances bacterial growth, as a result of which bacteria release inorganic nitrogen and other essential nutrients which can be utilized to increase bacterial biomass. These bacterial communities attract other microorganisms and act as food source for them.

6. Chemical and biological insecticides

To control agricultural insect pests, chemical pesticides are used continuously at high doses. The use of chemical pesticides has caused a number of issues, such as the resurgence of pests due to resistance and the destruction of natural predators. The large-scale use of pesticides not only costs a lot of money, but also causes residual effects, directly or indirectly, to animals, non-target soil fauna, and human beings. The increasing environmental concerns about the use of chemical pesticides and their decreasing availability reawakened the interest in biological control of insect pests. The primary purpose of implementing biological control programs is to reduce pest infestation so that consumers can purchase high-quality fruits, vegetables, and other plant-based products. Entomopathogenic nematodes have added a new dimension to Integrated Pest Management (IPM) strategies, due to which the interest in entomopathogenic nematodes in biological control of insect pests has dramatically increased and many nematologists have done classical work on them.

7. EPNs as bio-insecticides for control of insect pests

Entomopathogenic nematodes (EPNs) are widely distributed across the world and have been reported from a variety of natural and managed habitats and from different types of soils [93–95]. They are eco-friendly, safe to non-target organisms, have a great potential for reproduction [79], wide host range, easy to mass produce, capable of resisting under unfavorable conditions [96], easy to apply, and compatible with most insecticides [97]. After *Bacillus thuringiensis* [98], the second most utilized biological control agents are EPNs and they are used in high-value crops to reduce the population of insect pests. Biological control using EPNs offers a number of advantages over chemical control, particularly in the environmental and nutritional domains.

In biological control programs, the foraging strategies of EPN are useful to recommend an EPN species to be used as a biocontrol agent against insect pests. EPN species of genus *Steinernema* and *Heterorhabditis* are used as biocontrol agents because they have the ability to find and kill even deep-seated insects [99]. Moreover, they are fast-acting, cost-effective, and easy to handle. The EPN species with cruise foraging strategy are considered best to target immobile insect hosts in soil or in other cryptic habitats. Whereas, the ambush foragers are not suitable for this purpose [100]. However, more and more field studies suggest that this distinction might be oversimplified. Although there are significant differences in the normal foraging behavior of species, being classified as an ambusher does not preclude success against distant cryptic insects. This is exemplified by the ability of *S. carpocapsae*, which can parasitize the larvae and pupae of the large pine weevil *Hylobius abietis* that lives under the bark of tree roots [101]. EPNs are considered beneficial to farmers because of their ability to locate and kill insect pests. They are preferred as excellent biological control agents due to some important and fascinating attributes including locating above and below ground insect pests, symbiotic association with pathogenic bacteria, i.e., *Xenorhabdus* spp. with *Steinernema* spp. and *Photorhabdus* spp. with *Heterorhabditis* spp., great reproductive potential, and high pathogenicity against insect pests. The biocontrol potential of various EPN species has been studied against different insect pests and found to be very effective to control targeted insect pests. For example, *Heterorhabditis bacteriophora* was tested against the mediterranean fruit fly *Ceratitis capitata*, *Heterorhabditis zealandica* tested against citrus mealy bug *Planococcus citri* [102–104], and *Steinernema carpocapsae* against the large pine weevil *Hylobius abietis* [105].

8. The commercial success of EPNs

Numerous businesses and corporations are effectively manufacturing and marketing *Heterorhabditis* and *Steinernema* products in various nations around the globe. However, these nematodes are highly sensitive to environmental conditions (such as temperature, moisture, and sun exposure) which can influence their storage longevity and efficacy in the field. Nevertheless, various EPN species including *S. carpocapsae, S. feltiae, S. riobrave, S. scapterisci, S. kushidai, S. kraussei, H. bacteriophora, H. megidis, H. downesi,* and *Heterorhabditis indica* have been marketed as commercial products under different brand names. A list of insect pests commercially targeted with EPN species is provided in **Table 2**. On a global level, a number of species-specific EPNs are being utilized as biocontrol agents under various trade names such as Ecomask, Entonem, Scanmask, Heteromask, Nemasys, J-3 Max, Savior Weevil larvae, Lawn Patrol, Guardian, etc. and have shown the excellent remarkable results against the targeted insect pests. Commercially, the Bionema company (United Kingdom) and E-nema company (Europe) produce bioformulations using EPNs and earn millions of US dollars annually by selling these products. The significant interest and demand of EPN products have gradually increased in various countries such as the United States of America, Germany, Netherlands, Japan, Belgium, India, Austria, New Zealand,

Target Insect Pests		Crop(s) infested by Insect	Efficacious EPN Species
Common name	**Scientific name**		
Armyworms	*Spodoptera spp.*	Vegetables	*Steinernema carpocapsae, S. feltiae, S. riobrave*
Banana moth	*Opogona sachari*	Ornamentals	*Heterorhabditis bacteriophora, S. carpocapsae*
Banana root borer	*Cosmopolites sordidus*	Banana	*S. carpocapsae, S. feltiae, S. glaseri*
Billbug	*Sphenophorus spp.*	Turf	*H. bacteriophora, S. carpocapsae*
Black cutworm	*Agrotis ipsilon*	Turf, vegetables	*S. carpocapsae*
Black vine weevil	*Otiorhynchus sulcatus*	Berries, ornamentals	*H. bacteriophora, H. downesi, H. marelata, H. megidi, S. carpocapsae, S. glaseri*
Borers	*Synanthedon spp. and other sesiids*	Fruit trees & ornamentals	*H. bacteriophora, S. carpocapsae, S. feltiae*
Cat flea	*Ctenocephalides felis*	Home yard, turf	*S. carpocapsae*
Citrus root weevil	*Pachnaeus spp.*	Citrus, ornamentals	*S. riobrave, H. bacteriophora*
Codling moth	*Cydia pomonella*	Pome fruit	*S. carpocapsae, S. feltiae*
Corn earworm	*Helicoverpa zea*	Vegetables	*S. carpocapsae, S. feltiae, S. riobrave*
Corn rootworm	*Diabrotica spp.*	Vegetables	*H. bacteriophora, S. carpocapsae*
Cranberry girdler	*Chrysoteuchia topiaria*	Cranberries	*S. carpocapsae*
Crane fly	*Diptera: Tipulidae*	Turf	*S. carpocapsae*
Diaprepes root weevil	*Diaprepes abbreviatus*	Citrus, ornamentals	*H. bacteriophora, S. carpocapsae*
Fungus gnats	*Diptera: Sciaridae*	Mushrooms, greenhouse	*H. bacteriophora, S. feltiae*
Grape root borer	*Vitacea polistiformis*	Grapes	*H. zealandica, H. bacteriophora*
Iris borer	*Macronoctua onusta*	Iris	*H. bacteriophora, S. carpocapsae*
Large pine weevil	*Hylobius albietis*	Forest plantings	*H. downesi, S. carpocapsae*
Leafminers	*Liriomyza spp.*	Vegetables, ornamentals	*S. carpocapsae, S. feltiae*

Target Insect Pests		**Crop(s) infested by Insect**	**Efficacious EPN Species**
Common name	**Scientific name**		
Mole crickets	*Scapteriscus spp.*	Turf	*S. carpocapsae,* *S. riobrave,* *S. scapterisci*
Navel orangeworm	*Amyelois transitella*	Nut and fruit trees	*S. carpocapsae*
Plum curculio	*Conotrachelus nenuphar*	Fruit trees	*S. riobrave*
Scarab grubs	*Coleoptera: Scarabaeidae*	Turf, ornamentals	*H. bacteriophora,* *S. carpocapsae,* *S. glaseri,* *S. scarabaei,* *H. zealandica*
Shore flies	*Scatella spp.*	Ornamentals	*S. carpocapsae,* *S. feltiae*
Strawberry root weevil	*Otiorhynchus ovatus*	Berries	*H. marelata*
Small hive beetle	*Aethina tumida*	Bee hives	*H. indica,* *S. riobrave*
Sweetpotato weevil	*Cylas formicarius*	Sweet potato	*H. bacteriophora,* *S. carpocapsae,* *S. feltiae*

Table 2.
List of insect pests commercially targeted with EPNs.

Australia, United Kingdom, Canada Kenya, and Poland. In 1980s, non-native EPN species *S. carpocapsae* and *H. bacteriophora* were mass produced by Ecomax company under the trade names, Green Commandos and Soil Commandos, respectively but the results under field conditions provided by these products were inconsistent in India, maybe due to their poor adaptability to Indian environmental conditions or due to their production or formulation problems. Later, both of these products were discontinued and also withdrawn from the market at the same time. Therefore, the best and most effective strategy is to utilize indigenous species that are present locally which may provide outstanding results.

9. Conclusion

In conclusion, nematodes associated with insects are considered beneficial organisms in agriculture and gardening because some of the nematodes have the ability to control the insect pest populations by destroying their soil-dwelling larvae, and free-living nematodes have the ability to improve the soil structure and growing process of plants by providing the nutrients to the soil. The tripartite interaction between Nematodes-Bacteria-Insects demonstrates a remarkable level of adaptation and evolutionary cooperation. EPNs from the families Steinernematidae and Heterorhabditidae have been recognized as powerful biological control agents due to their synergistic combination with specific bacteria, mainly from the genera *Xenorhabdus* and *Photorhabdus.* The wide distribution of these nematodes and their ecological adaptability makes them potential bio-agents which can be utilized in integrated pest management strategies to manage various soil-inhabiting insect pests in various agricultural systems.

Author details

Heena Khan[1*], Nitesh Kumar[2] and Ashok Kumar Chaubey[3]

1 Department of Zoology, Chaudhary Charan Singh University, Meerut, India

2 Department of Zoology, S.G.T.B.S. Govt. Degree College, Bilaspur, Rampur, India

3 Chaudhary Charan Singh University, Meerut, India

*Address all correspondence to: heena.zoology@gmail.com

References

[1] Dhaliwal GS, Dhawan AK, Singh R. Biodiversity and ecological agriculture: Issues and perspectives. Indian Journal of Ecology. 2007;**34**:100-109

[2] Dhaliwal GS, Jindal V, Mohindru B. Crop losses due to insect pests: Global and Indian scenario. Indian Journal of Entomology. 2015;**77**(2):165-168. DOI: 10.5958/0974-8172.2015.00033.4

[3] Giblin-Davis RM, Kanzaki N, Davies KA. Nematodes that ride insects: Unforeseen consequences of arriving species. The Florida Entomologist. 2013;**96**(3):770-780. Available from: http://www.jstor.org/stable/23609386

[4] Mrácek Z. Use of entomopathogenic nematodes (EPANS) in biological control. In: Upadhyay RK, editor. Advances in Microbial Control of Insect Pests. USA: Springer; 2008. pp. 235-264

[5] Blouin M, Liu J, Berry R. Life cycle variation and the genetic structure of nematode populations. Heredity. 1999;**83**:253-259. DOI: 10.1038/sj.hdy.6885420

[6] Bedding RA, Akhurst RJ. Geographical distribution and host preferences of *Deladenus* species (Nematoda: Neotylenchidae) parasitic in siricid woodwasps and associated hymenopterous parasitoids. Nematologica. 1978;**24**(3):286-294. DOI: 10.1163/187529278X00254

[7] Heena, Kumar N, Chaubey AK. Redescription and molecular characterization of *Panagrellus ceylonensis* (Nematoda: Rhabditida, Panagrolaimidae) from India. Zootaxa. 2023;**5383**(4):491-513. DOI: 10.11646/zootaxa.5383.4.4

[8] Richardson PN, Hesling JJ, Riding IL. Life cycle and description of *Howardula husseyi* n. sp. (Tylenchida: Allantonematidae), a nematode parasite of the mushroom phorid *Megaselia halterata* (Diptera: Phoridae). Nematologica. 1977;**23**:217-231

[9] Kuris AM, Lafferty KD. Parasite–host modelling meets reality: Adaptive peaks and their ecological attributes. In: Poulin R, Morand S, Skorping A, editors. Evolutionary Biology of Host-Parasite Relationships: Theory Meets Reality. Amsterdam: Elsevier Science; 2000. pp. 9-26

[10] Lafferty KD, Kuris AM. Trophic strategies, animal diversity and body size. Trends in Ecology and Evolution. 2002;**17**:507-513

[11] Poulin R. The many roads to parasitism: A tale of convergence. Advances in Parasitology. 2011;**74**:1-40

[12] Poulin R, Randhawa HS. Evolution of parasitism along convergent lines: From ecology to genomics. Parasitology. 2013;**142**(Suppl 1):S6-S15. DOI: 10.1017/S0031182013001674

[13] Giblin-Davis RM. Borers of palms. In: Howard FW, Moore D, Giblin-Davis RM, Abad R, editors. Insects on Palms. Wallingford, UK: CABI publishing; 2001. pp. 267-304. DOI: 10.1079/9780851993263.0267

[14] Kanzaki N, Giblin-Davis RM. Pine wilt and red ring, lethal plant diseases caused by insect-mediated *Bursaphelenchus* nematodes. In: Brown JK, editor. Vectormediated Transmission of Plant Pathogens. St Paul, Minnesota, USA: APS Press; 2016. pp. 87-107. DOI: 10.1094/9780890545355.007

[15] Abebe E, Bonner K, Gray V, Thomas WK. Response to bugs don't

make worms kill. Journal of Experimental Biology. 2011;**214**:1053-1054. DOI: 10.1242/jeb.053827

[16] Lephoto TE, Featherston J, Gray VM. Draft whole-genome sequence of *Serratia* sp. strain TEL, associated with *Oscheius* sp. TEL-2014 (Nematoda: Rhabditidae) isolated from a grassland in South Africa. Genome Announcements. 2015;**3**(4):e00747-e00715. DOI: 10.1128/genomeA.00747-15

[17] Loulou A, Mastore M, Caramella S, Bhat AH, Brivio MF, Machado RAR, et al. Entomopathogenic potential of bacteria associated with soil-borne nematodes and insect immune responses to their infection. PLoS One. 2023;**18**(1):e0280675. DOI: 10.1371/journal. pone.0280675

[18] Doucet ME, Doucet MMA. Contribution to the knowledge of *Distolabrellus veechi* Anderson, 1983 (Nemata: Rhabditida). Fundamental and Applied Nematology. 1992;**15**:413-420

[19] Ali M, Wahab A, El-Kifel AH. Nematodes associated with Coleoptera species in Egypt I. Parasitologia Hungarica. 1972;**5**:177-202

[20] Boström S. Descriptions and morphological variability of three populations of Panagrolaimus Fuchs, 1930 (Nematoda: Panagrolaimidae). Nematologica. 1989;**34**:144-155

[21] Geraert E, Mracek Z, Sudhaus W. The facultative insect parasite Prislionchus Iherilieri (Syn. P. Pseudolherilien) (Nematoda: Diplogasterida). Nematologica. 1990;**35**:133-141

[22] Tsai MH, Tang LC, Hou RF. The bacterium associated with the entomopathogenic nematode Steinernema abbasi (Nematoda: Stein ernematidae) isolated from Taiwan. Journal of Invertebrate Pathology. 2008;**99**:242-245

[23] Kuwata R, Qiu L-h, Wang W, Harada Y, Yoshida M, Kondo E, et al. *Xenorhabdus ishibashii* sp. nov., isolated from the entomopathogenic nematode Steinernema aciari. International Journal of Systematic and Evolutionary Microbiology. 2013;**63**:1690-1695

[24] Akhurst R, Boemare N. A numerical taxonomic study of the genus *Xenorhabdus* (Enterobacteriaceae) and proposed elevation of the sub species of *X. Nematophilus* to species. Microbiology. 1988;**134**:1835-1845

[25] Stock SP, Bana AL, Darwish R, Katbeh A. Diversity and distribution of entomopathogenic nematodes (Nematoda: Steinernematidae, Heterorhabditidae) and their bacterial symbionts in Jordan. Journal of Invertebrate Pathology. 2008;**98**:228-234. DOI: 10.1016/j.jip.2008.01.003

[26] Tailliez P, Pagès S, Ginibre N, Boemare N. New insight into diversity in the genus *Xenorhabdus*, including the description of ten novel species. International Journal of Systematic and Evolutionary Microbiology. 2006;**56**:2805-2818. DOI: 10.1099/ijs.0.64287-0

[27] Tailliez P, Pagès S, Edgington S, Tymo LM, Buddie AG. Description of *Xenorhabdus magdalenensis* sp. nov., the symbiotic bacterium associated with *Steinernema australe*. International Journal of Systematic and Evolutionary Microbiology. 2012;**62**:1761-1765. DOI: 10.1099/ijs.0.034322-0

[28] Lengyel K, Lang E, Fodor A, Szállás E, Schumann P, Stackebrandt E. Description of four novel species of *Xenorhabdus*, family Enterobacteriaceae: *Xenorhabdus budapestensis* sp. nov.,

Xenorhabdus ehlersii sp. nov., *Xenorhabdus innexi* sp. nov., and *Xenorhabdus szentirmaii* sp. nov. Systematic and Applied Microbiology. 2005;**28**:115-122. DOI: 10.1016/j.syapm.2004.10.004

[29] Cimen H, Půža V, Nermuť J, Hatting J, Ramakuwela T, Faktorova L, et al. *Steinernema beitlechemi* n. sp., a new entomopathogenic nema tode (Nematoda: Steinernematidae) from South Africa. Nematology. 2016;**18**:439-453

[30] Cimen H, Půža V, Nermu Ť J, Hatting J, Ramakuwela T, Hazir S. *Steinernema biddulphi* n. sp., a new entomopathogenic nematode (Nema Toda: Steiner-nematidae) from South Africa. Journal of Nematology. 2016;**48**:148

[31] Yang J et al. An insecticidal protein from *Xenorhabdus budapestensis* that results in prophenoloxidase activation in the wax moth, *galleria mellonella*. Journal of Invertebrate Pathology. 2012;**110**:60-67

[32] Fischer-Le Saux M, Arteaga-Hernandez E, Mracek Z, Boemare N. The bac terial symbiont *Xenorhabdus poinarii* (Enterobacteriaceae) is harbored by two phylogenetic related host nematodes: The entomopathogenic species *Steinernema cubanum* and *Steinernema glaseri* (Nematoda: Steinernematidae). FEMS Microbiology Ecology. 1999;**29**:149-157

[33] Martens EC, Goodrich-Blair H. The *Steinernema carpocapsae* intestinal vesicle contains a subcellular structure with which *Xenorhabdus nematophila* associates during colonization initiation. Cellular Microbiology. 2005;**7**:1723-1735

[34] Stokwe NF, Malan AP, Knoetze NKB, R and Tiedt L. *Steinernema citrae* n. sp. (Rhabditida: Steinernematidae), a new entomopathogenic nematode from South Africa. Nematology. 2011;**13**:569-587

[35] Kämpfer P, Tobias NJ, Ke LP, Bode HB, Glaeser SP. *Xenorhabdus thuongxuanensis* sp. nov. and *Xenorhabdus eapokensis* sp. nov., isolated from Steinernema species. International Journal of Systematic and Evolutionary Microbiology. 2017;**67**:1107-1114

[36] Abate BA, Slippers B, Wingfield MJ, Malan AP, Hurley BP. Diversity of entomopathogenic nematodes and their symbiotic bacteria in south African plantations and indigenous forests. Nematology. 2018;**20**:355-371

[37] Ehlers RU, Wulff A, Peters A. Pathogenicity of axenic *Steinernema feltiae*, *Xenorhabdus bovienii,* and the bacto-helminthic complex to larvae of *Tipula oleracea* (Diptera) and *galleria mellonella* (Lepidoptera). Journal of Invertebrate Pathology. 1997;**69**(3):212-217

[38] Akhurst RJ. Taxonomic study of *Xenorhabdus*, a genus of bacteria symbiotically associated with insect pathogenic nematodes. International Journal of Systematic and Evolutionary Microbiology. 1983;**33**:38-45

[39] Tarasco E, Santiago Alvarez C, Triggiani O, Quesada ME. Laboratory studies on the competition for insect haemocoel between *Beauveria bassiana* and *Steinernema ichnusae* recovered in the same ecological niche. Biocontrol Science and Technology. 2011;**21**:693-704

[40] Dreyer J, Rautenbach M, Booysen E, Van Staden A, Deane S, Dicks L. *Xenorhabdus khoisanae* SB10 produces Lys-rich PAX lipopeptides and a Xenocoumacin in its antimicrobial complex. BMC Microbiology. 2019;**19**:1-11

[41] Ferreira T, Van Reenen CA, Endo A, Spröer C, Malan AP, Dicks LM. Description of *Xenorhabdus khoisanae* sp. nov., the symbiont of the entomopathogenic nematode

Steinernema khoisanae. International Journal of Systematic and Evolutionary Microbiology. 2013;**63**:3220-3224

[42] Nishimura Y, Hagiwara A, Suzuki T, Yamanaka S. *Xenorhabdus japonicus* sp. nov. associated with the nematode Steinernema kushidai. World Journal of Microbiology and Biotechnology. 1994;**10**:207-210

[43] Akhurst R. A *Xenorhabdus* sp. (Eubacteriales: Enterobacteriaceae) sym biotically associated with *Steinernema kraussei* (Nematoda: Steinernematidae). Revue Nematology. 1982;5:277-280

[44] Baniya A, DiGennaro P. Genome announcement of *Steinernema khuongi* and its associated symbiont from Florida. G3 (Bethesda). 2021;**11**(4):jkab053. DOI: 10.1093/g3journal/jkab053

[45] Özdemir E, Bayram Ş, Susurluk IA. First record of the entomopathogenic nematode *Steinernema litorale* (Filipjev) (Rhabditida: Steinernematidae) and its symbiotic bacterium from Turkey, and its efficacy capability. Insects. 2020;**11**:144

[46] Kang S-J, Han S-C, Yi Y-K, Kim Y-G. Identification of an entomopathogenic nematode, *Steinernema monticolum*, and its novel symbiotic bacterium. In: Proceedings of the Korean Society of Applied Entomology Conference. Korea: Korean Society of Applied Entomology; 2003. p. 83

[47] Sajnaga E, Kazimierczak W, Skowronek M, Lis M, Skrzypek T, Waśko A. *Steinernema poinari* (Nematoda: Steinernematidae): A new symbiotic host of entomopathogenic bacteria *Xenorhabdus bovienii*. Archives of Microbiology. 2018;**200**:1307-1316

[48] Tailliez P, Laroui C, Ginibre N, Paule A, Pagès S, Boemare N. Phylogeny of Photorhabdus and *Xenorhabdus* based on universally conserved protein-coding sequences and implications for the taxonomy of these two genera. Proposal of new taxa: *X. Vietnamensis* sp. nov., *P. Luminescens* subsp. *caribbeanensis* subsp. nov., *P. Luminescens* subsp. *hainanensis* subsp. nov., *P. Temperata* subsp. khanii subsp. nov., *P. Temperata* subsp. *tasmaniensis* subsp. nov., and the reclassification of *P. Luminescens* subsp. *thracensis* as *P. Temperata* subsp. *thracensis* comb. nov. International Journal of Systematic and Evolutionary Microbiology. 2010;**60**:1921-1937

[49] Kim IH, Aryal SK, Aghai DT, Casanova-Torres ÁM, Hillman K, Kozuch MP, et al. The insect pathogenic bacterium *Xenorhabdus innexi* has attenuated virulence in multiple insect model hosts yet encodes a potent mosquitocidal toxin. BMC Genomics. 2017;**18**(1):1-25

[50] Ardpairin J et al. A survey of entomopathogenic nematodes and their symbiotic bacteria in agricultural areas of northern Thailand. Journal of Helminthology. 2020;**94**:1-11

[51] Bhat AH, Istkhar AKC, Půža V, San-Blas E. First report and comparative study of Steinernema surkhetense (Rhabditida: Steinernematidae) and its symbiont bacteria from subcontinental India. Journal of Nematology. 2017;**49**:92

[52] Gorgadze O, Lortkhipanidze M, Ogier J-C, Tailliez P, Burjanadze M. *Steinernema tbilisiensis* sp. n. (Nematoda: Steinernematidae)—A new species of entomopathogenic nematode from Georgia. Journal of Agricultural Science and Technology. 2015;**5**:264-276

[53] Castaneda-Alvarez C, Prodan S, Zamorano A, San-Blas E, Aballay E. *Xenorhabdus lircayensis* sp. nov., the symbiotic bacterium associated with the entomopathogenic nematode

Steinernema unicornum. International Journal of Systematic and Evolutionary Microbiology. 2021;**71**:005151. DOI: 10.1099/ijsem.0.005151

[54] Clausi M, Longo A, Rappazzo G, Tarasco E, Vinciguerra MT. *Steinernema vulcanicum* n. sp. (Rhabditida: Steinernematidae), a new entomopathogenic nematode species from Sicily (Italy). Nematology. 2011;**13**:409-423

[55] Tamiru T, Waeyenberge L, Hailu T, Ehlers R-U, Půža V, Mráček Z. *Steinernema ethiopiense* sp. n. (Rhabditida: Steinernematidae), a new entomopathogenic nematode from Ethiopia. Nematology. 2012;**14**:741-757

[56] Machado RAR, Bruno P, Arce CCM, Liechti N, Köhler A, Bernal J, et al. *Photorhabdus khanii* subsp. *guanajuatensis* subsp. nov., isolated from *Heterorhabditis atacamensis*, and *Photorhabdus luminescens* subsp. *mexicana* subsp. nov., isolated from *Heterorhabditis mexicana* entomopathogenic nematodes. International Journal of Systematic and Evolutionary Microbiology. 2019;**69**(3):652-661. DOI: 10.1099/ijsem.0.003154. Epub 2019 Jan 28

[57] Boemare N, Akhurst R, Mourant R. DNA relatedness between *Xenorhabdus* spp. (Enterobacteriaceae), symbiotic bacteria of entomopathogenic nematodes, and a proposal to transfer *Xenorhabdus luminescens* to a new genus, *Photorhabdus* gen. Nov. International Journal of Systematic and Evolutionary Microbiology. 1993;**43**:249-255

[58] Machado RAR, Wüthrich D, Kuhnert P, Arce CCM, Thönen L, Ruiz C, et al. Whole-genome-based revisit of *Photorhabdus* phylogeny: Proposal for the elevation of most *Photorhabdus* subspecies to the species level and description of one novel species *Photorhabdus bodei* sp. nov., and one novel subspecies *Photorhabdus laumondii* subsp. *clarkei* subsp. nov. International Journal of Systematic and Evolutionary Microbiology. 2018 Aug;**68**(8):2664-2681. DOI: 10.1099/ijsem.0.002820. Epub 2018 Jun 7

[59] Machado RAR, Muller A, Ghazal SM, Thanwisai A, Pagès S, Bode HB, et al. *Photorhabdus heterorhabditis* subsp. *aluminescens* subsp. nov., *Photorhabdus heterorhabditis* subsp. *heterorhabditis* subsp. nov., *Photorhabdus australis* subsp. *thailandensis* subsp. nov., *Photorhabdus australis* subsp. *australis* subsp. nov., and *Photorhabdus aegyptia* sp. nov. isolated from *Heterorhabditis* entomopathogenic nematodes. International Journal of Systematic and Evolutionary Microbiology. 2021;**71**:4610. DOI: 10.1099/ijsem.0.004610

[60] Glaeser SP, Tobias NJ, Thanwisai A, Chantratita N, Bode HB, Kämpfer P. *Photorhabdus luminescens* subsp. *namnaonensis* subsp. nov., isolated from *Heterorhabditis baujardi* nematodes. International Journal of Systematic and Evolutionary Microbiology. 2017;**67**:1046-1051

[61] Bhat AH, Machado RAR, Abolafia J, Ruiz-Cuenca AN, Askary TH, Ameen F, et al. Taxonomic and molecular characterization of a new entomopathogenic nematode species, *Heterorhabditis casmirica* n. sp., and whole genome sequencing of its associated bacterial symbiont. Parasites & Vectors. 2023;**16**(1):383. DOI: 10.1186/s13071-023-05990-z. Erratum in: Parasit Vectors. 2023 Nov 27;16(1):439. DOI: 10.1186/s13071-023-06060-0

[62] Blackburn D, Crawford B, Shapiro-Ilan DI, Adams BJ. Environmental drivers of trait changes in *Photorhabdus luminescens*. Biological Control. 2016;**92**:145-152

[63] Plichta KL, Joyce SA, Clarke D, Waterfield N, Stock SP. *Heterorhabditis gerrardi* n. sp. (Nematoda: Heterorhabditidae): The hidden host of *Photorhabdus asymbiotica* (Enterobacteriaceae: Γ-Proteobacteria). Journal of Helminthology. 2009;**83**(4):309-320

[64] Akhurst RJ, Boemare NE, Janssen PH, Peel MM, Alfredson DA, Beard CE. Taxonomy of Australian clinical isolates of the genus *Photorhabdus* and proposal of *Photorhabdus asymbiotica* subsp. *asymbiotica* subsp. nov. and *P. asymbiotica* subsp. *australis* subsp. nov. International Journal of Systematic and Evolutionary Microbiology. 2004;**54**(4):1301-1310

[65] Tóth T, Lakatos T. *Photorhabdus temperata* subsp. cinerea subsp. nov., isolated from *Heterorhabditis* nematodes. International Journal of Systematic and Evolutionary Microbiology. 2008;**58**:2579-2581

[66] Geldenhuys J, Malan A, Dicks L. First report of the isolation of the symbiotic bacterium *Photorhabdus luminescens* subsp. *laumondii* associated with *Heterorhabditis safricana* from South Africa. Current Microbiology. 2016;**73**:790-795

[67] Orozco RA, Hill T, Stock SP. Characterization and phylogenetic relationships of *Photorhabdus luminescens* subsp. *sonorensis* (γ-Proteobacteria: Enterobacteriaceae), the bacterial symbiont of the entomopathogenic nematode *Heterorhabditis sonorensis* (Nematoda: Heterorhabditidae). Current Microbiology. 2013;**66**:30-39

[68] Ferreira T, van Reenen CA, Endo A, Tailliez P, Pagès S, Spröer C, et al. *Photorhabdus heterorhabditis* sp. nov., a symbiont of the entomopathogenic nematode *Heterorhabditis zealandica*. International Journal of Systematic and Evolutionary Microbiology. 2014 May;**64**(Pt 5):1540-1545. DOI: 10.1099/ijs.0.059840-0. Epub 2014 Jan 29

[69] Machado RAR, Somvanshi VS, Muller A, Kushwah J, Bhat CG. *Photorhabdus hindustanensis* sp. nov., *Photorhabdus akhurstii* subsp. *akhurstii* subsp. nov., and *Photorhabdus akhurstii* subsp. *bharatensis* subsp. nov., isolated from Heterorhabditis entomopathogenic nematodes. International Journal of Systematic and Evolutionary Microbiology. 2021;**71**:4998-5006. DOI: 10.1099/ijsem.0.004998

[70] Zhang C-X, Yang S-Y, Xu M-X, Sun J, Liu H, Liu J-R, et al. *Serratia nematodiphila* sp. nov., associated symbiotically with the entomopathogenic nematode *Heterorhabditidoides chongmingensis* (Rhabditida: Rhabditidae). International Journal of Systematic and Evolutionary Microbiology. 2009;**59**:1603-1608. DOI: 10.1099/ijs.0. 65718-0

[71] Zhan C, Chen L, Guo D, Sun J, Duan Y, Zhang P, et al. An intestinal symbiotic bacterial strain of *Oscheius chongmingensis* modulates host viability at both global and post-transcriptional levels. International Journal of Molecular Sciences. 2022;**23**:14692. DOI: 10.3390/ijms232314692

[72] Fu J-R, Liu Q-Z. Evaluation and entomopathogenicity of gut bacteria associated with dauer juveniles of *Oscheius chongmingensis* (Nematoda: Rhabditidae). Microbiology Open. 2019;**8**:e823. DOI: 10.1002/mbo3.823

[73] Torres-Barragan A, Suazo A, Buhler WG, Cardoza YJ. Studies on the entomopathogenicity and bacterial associates of the nematode Oscheius carolinensis. Biological Control.

2011;**59**(2):123-129. DOI: 10.1016/j.biocontrol.2011.05.020

[74] Zhou G, Yang H, Wang F, Bao H, Wang G, Hou X, et al. Oscheius microvilli n. sp. (Nematoda: Rhabditidae): A facultatively pathogenic nematode from Chongming Island, China. Journal of Nematology. 2017;**49**:33-41. DOI: 10.21307/jofnem 2017-044

[75] Park HW, Kim YO, Ha J, Youn SH, Kim HH, Bilgrami AL, et al. Effects of associated bacteria on the pathogenicity and reproduction of the insect-parasitic nematode *Rhabditis blumi* (Nematoda: Rhabditida). Canadian Journal of Microbiology. 2011;**57**:750-758

[76] Steiner G. Aplectana kraussei n. sp., eine in der Blattwespe Lyda sp. parasitierende Nematoden form, nebst Bemerkungen über das Seitenorgan der parasitis chen Nematoden. Zentralblatt für Bakteriologie, Parasitenkunde, Infektions krankheiten und Hygiene, Abteilung II. 1923;**59**:14-18

[77] Poinar GO Jr. Taxonomy and biology of Steinernematidae and Heterorhabditidae. In: Gaugler R, Kaya HK, editors. Entomopathogenic Nematodes in Biological Control. Boca Raton, FL, USA: CRC Press; 1990. pp. 23-62

[78] Koppenhöfer AM. Nematodes. In: Lacey LA, Kaya HK, editors. Field Manual of Techniques in Invertebrate Pathology: Application and Evaluation of Pathogens for Control of Insects and Other Invertebrate Pests. 2nd ed. Dordrecht: Springer; 2007. pp. 249-264

[79] Kaya H, Gaugler R. Entomopathogenic nematodes. Annual Review of Entomology. 1993;**38**:181-206

[80] Thanwisai A, Tandhavanant S, Saiprom N, Waterfield NR, Ke P, Bode HB, et al. Diversity of *Xenorhabdus* and *Photorhabdus* spp. and their symbiotic entomopathogenic nematodes from Thailand. PLoS One. 2012;**7**(9):e43835

[81] Yooyangket T, Muangpat P, Polseela R, Tandhavanant S, Thanwisai A, Vitta A. Identification of entomopathogenic nematodes and symbiotic bacteria from Nam Nao National Park in Thailand and larvicidal activity of symbiotic bacteria against *Aedes aegypti* and *Aedes albopictus*. PLoS One. 2018;**13**:e0195681

[82] Gozel U, Gozel C. Entomopathogenic nematodes in pest management. In: Gill H, editor. Integrated Pest Management (IPM): Environmentally Sound Pest Management. London, UK, London, UK: IntechOpen; 2016. pp. 55-69

[83] Zhang C, Liu J, Xu M, Sun J, Yang S, An X, et al. *Heterorhabditidoides chongmingensis* gen. Nov., sp. nov. (Rhabditida: Rhabditidae), a novel member of the entomopathogenic nematodes. Journal of Invertebrate Pathology. 2008;**98**(2):153-168. DOI: 10.1016/j.jip.2008.02.011

[84] Ye W, Torres-Barragan A, Cardoza YJ. *Oscheius carolinensis* n. sp. (Nematoda: Rhabditidae), a potential entomopathogenic nematode from vermicompost. Nematology. 2010;**12**(1):121-135

[85] Pervez R, Eapen SJ, Devasahayam S, Jacob TK. A new species of entomopathogenic nematode *Oscheius gingeri* sp. n. (Nematoda: Rhabditidae) from ginger rhizosphere. Archives of Phytopathology and Plant Protection. 2013;**46**:526-535

[86] Torrini G, Mazza G, Carletti B, Benvenuti C, Roversi PF, Fanelli E, et al. *Oscheius onirici* sp. n. (Nematoda: Rhabditidae): A new entomopathogenic

nematode from an Italian cave. Zootaxa. 2015;**3937**(3):533-548. DOI: 10.11646/zootaxa.3937.3.6

[87] Zhang K, Baiocchi T, Lu D, Chang DZ, Dillman AR. Differentiating between scavengers and entomopathogenic nematodes: Which is *Oscheius chongmingensis*? Journal of Invertebrate Pathology. 2019;**167**:107245. DOI: 10.1016/j.jip.2019.107245

[88] Lewis EE, Hazir S, Hodson A, Gulcu B. Trophic relationships of entomopathogenic nematodes in agricultural habitats. In: Nematode Pathogenesis of Insects and Other Pests: Ecology and Applied Technologies for Sustainable Plant and Crop Protection. Springer International Publishing; 2015. pp. 139-163. DOI: 10.1007/978-3-319-18266-7_5

[89] Dillman AR, Chaston JM, Adams BJ, Ciche TA, Goodrich-Blair H, Sternberg PW. An entomopathogenic nematode by any other name. PLoS Pathogens. 2012;**8**:e1002527. DOI: 10.1371/journal.ppat.1002527

[90] San-Blas E, Gowen SR. Facultative scavenging as a survival strategy of entomopathogenic nematodes. International Journal for Parasitology. 2008;**38**(1):85-91. DOI: 10.1016/j.ijpara.2007.06.003

[91] Půža V, Mráček Z. Does scavenging extend the host range of entomopathogenic nematodes (Nematoda: Steinernematidae)? Journal of Invertebrate Pathology. 2010;**104**(1):1-3. DOI: 10.1016/j.jip.2010.01.002

[92] Baiocchi T, Lee G, Choe and Dillman AR. Host seeking parasitic nematodes use specific odors to assess host resources. Scientific Reports. 2017;**7**:6270. DOI: 10.1038/s41598-017-06620-2

[93] Hominick WM. Biogeography. In: Gaugler R, editor. Entomopathogenic Nematology. Vol. 1. New York: CABI publishing; 2002. pp. 115-143

[94] Adams BJ, Fodor A, Koppenhöfer HS, Stackebrandt E, Stock SP, Klein MG. Biodiversity and systematics of nematode–bacterium entomopathogens. Biological Control. 2006;**38**:4-21

[95] Adams BJ, Peat SM, Dillman AR. Phylogeny and evolution. In: Nguyen KB, Hunt DJ, editors. Entomopathogenic Nematodes: Systematics, Phylogeny and Bacterial Symbionts. Leiden-Boston: Brill; 2007. pp. 693-733

[96] Askary TH, Ahmad MA. Entomopathogenic nematodes: Mass production, formulation and application. In: Abd-Elgawad MM, Askary TH, Coupland J, editors. Biocontrol Agents: Entomopathogenic and Slug Parasitic Nematodes. UK: CABI Publishing; 2017. pp. 261-286

[97] Chen S. Infectivity and Persistence of Entomopathogenic Nematodes Used to Control the Cabbage Root Fly *Delia radicum* [Thesis]. Belgium: University of Ghent; 2003. p. 135

[98] Ehlers RU. Current and future use of nematodes in biocontrol: Practice and commercial aspects with regard to regulatory policy issues. Biocontrol Science and Technology. 1996;**6**:303-316

[99] Bedding RA, Miller LA. Disinfesting blackcurrant cuttings of *Synanthedon tipuliformis*, using the insect parasitic nematode, Neoaplectana bibionis. Environmental Entomology. 1981;**10**(4):449-453

[100] Gaugler R, Lewis E, Stuart RJ. Ecology in the service of biological control: The case of entomopathogenic nematodes. Oecologia. 1997;**109**:483-489. DOI: 10.1007/s004420050108

[101] Griffin CT. Perspectives on the behaviour of entomopathogenic nematodes from dispersal to reproduction: Traits contributing to nematode fitness and biocontrol efficacy. Journal of Nematology. 2012;**44**(2):177-184

[102] Van Niekerk S, Malan AP. Potential of south African entomopathogenic nematodes (Heterorhabditidae and Steinernematidae) for control of the citrus mealybug, *Planococcus citri* (Pseudococcidae). Journal of Invertebrate Pathology. 2012;**111**(2):166-174. DOI: 10.1016/j.jip. 2012.07.023

[103] Van Niekerk S, Malan AP. Adjuvants to improve control of *Planococcus citri* (Hemiptera: Pseudococcidae) using entomopathogenic nematodes (Heterorhabditidae and Steinernematidae). Journal of Helminthology. 2013;**89**(2):189-195. DOI: 10.1017/S0022149X13000771

[104] Van Niekerk S, Malan AP. Evaluating the efficacy of a polymer-surfactant formulation to improve control of the citrus mealybug, *Planococcus citri* (Risso) (Hemiptera: Pseudococcidae), using entomopathogenic nematodes under simulated natural conditions. African Plant Protection. 2014;**17**:1-8. DOI: 10.10520/EJC146102

[105] Brixey JM, Moore R, Milner AD. Effect of entomopathogenic nematode (*Steinernema carpocapsae* Weiser) application technique on the efficacy and distribution of infection of the large pine weevil (*Hylobius abietis* L.) in stumps of Sitka spruce (*Picea sitchensis* Carr.) created at different times. Forest Ecology and Management. 2006;**226**(1-3):161-172. DOI: 10.1016/j.foreco.2006.01.044. ISSN-0378-1127

Chapter 6

Expanding Host Range of Invasive Pest *Scirtothrips dorsalis* Hood in the U.S.A.

Sriyanka Lahiri and Chastity Perry

Abstract

The invasive thrips pest *Scirtothrips dorsalis* Hood has emerged as an important economic pest of strawberries and other specialty crops in FL, U.S.A. The alternate plant hosts being utilized by this pest during non-crop seasons is unknown and this information can benefit integrated pest management of *S. dorsalis*. Therefore, the objective of this two-year field study was to identify reproductive hosts of *S. dorsalis* in the woodland borders of strawberry fields. The hypothesis was that *S. dorsalis* has a weak dispersal capability and shows aggregated distribution on a range of plant families, therefore, several wild plants may be acting as hosts around crop farms. By conducting data collection in multiple commercial strawberry fields, several wild plant hosts were confirmed as reproductive hosts of *S. dorsalis*. These host plants include muscadine grape, Laurel cherry, sugarberry, Spanish needles, Peruvian water primrose, ragweed, sweetgum, and water oak. The temporal distribution of *S. dorsalis* on these plants across the commercial field sites on a monthly basis was also reported in this study. This finding can be applied for targeted biological control efforts, both in the case of augmentative and conservation biological control of *S. dorsalis.*

Keywords: chili thrips, woodland, reproductive host, biological control, strawberry

1. Introduction

Woodland and weedy areas around farms have played an important role in the establishment of certain invasive pests in the U.S.A. due to the availability of alternate plant hosts and refuge sites during non-cropping periods of the year. Examples include kudzu bug, *Megacopta cribraria* F. (Hemiptera: Plataspidae), and brown marmorated stink bug, *Halyomorpha halys* Stal (Hemiptera: Pentatomidae) [1, 2]. However, it is important to note that these landscapes have also provided an opportunity for native natural enemies to suppress the invasive pests. *Megacopta cribraria* suffered severe infection from entomopathogenic fungal strains while overwintering among the kudzu vines in woodlands in the southeastern U.S.A. [1]. Since 2015, the chili thrips *Scirtothrips dorsalis* Hood (Thysanoptera: Thripidae) has emerged as a severely damaging pest of strawberry, *Fragaria x ananassa* Duchesne

(Rosaceae) in the state of Florida (**Figure 1**). As is the case among thrips, *S. dorsalis* lays its eggs inside leaf tissues (**Figure 2**). Endemic to the southeast Asian region, *S. dorsalis* is known to have a broad host range. Regionally, Florida is the only state in the U.S.A. that produces strawberries for the fresh food market during the winter months from November to March, with an estimated annual revenue of $511 million [3]. As the second largest producer of strawberry in the U.S.A., fruit losses from feeding damage caused by *S. dorsalis* causes economic hardship

Figure 1.
Female Scirtothrips dorsalis *Hood. The adult, and first and second instar larvae are feeding stages with a preference for young foliage and fruits. Image credit: Joseph Montemayor, UF/IFAS GCREC.*

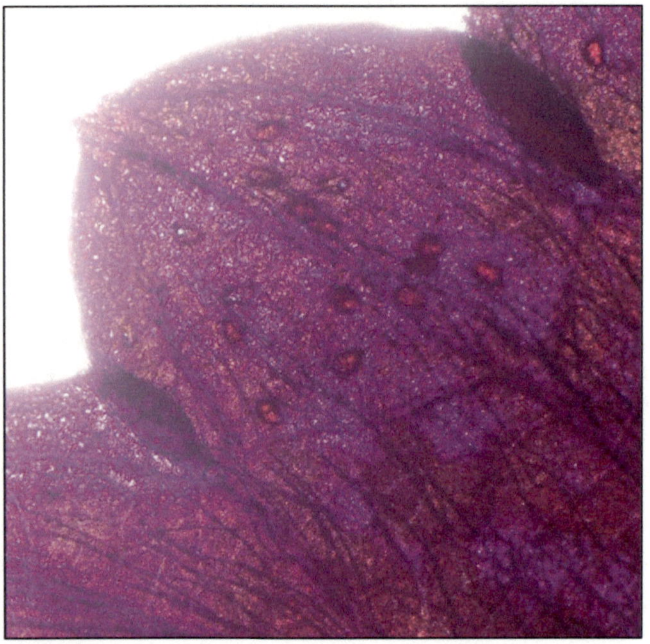

Figure 2.
Eggs of Scirtothrips dorsalis *Hood visible as light red grains inside strawberry leaf tissue. Image credit: Lovely Adhikary, UF/IFAS GCREC.*

to the industry. *Scirtothrips dorsalis* is a foliage feeder that colonizes strawberry transplants when they have young foliage, causing necrosis of leaves and abortion of flowers. Fruits, if present, also suffer feeding losses from *S. dorsalis*, that tend to hide under the fruit calyx and cause bronzing and cracking of fruits. Plants that are left untreated with any chemical insecticide may produce 1/8th marketable fruit yield compared to those plants that are treated with spinosyn based insecticides. Additionally, the management of *S. dorsalis* is difficult as it shows development of resistance to chemical insecticides such as spinosyns, diamides, acetamiprid, and pyrethroid [4]. Although an urgent development of an integrated pest management approach is needed to manage *S. dorsalis*, knowledge is still limited regarding its expanding host range in the U.S.A.

Since its arrival in the continental U.S.A., roses (Rosaceae), peppers (Solanaceae), strawberry (Rosaceae), blueberry (Ericaceae), and citrus (Rutaceae) have been reported as crop hosts in the state of Florida, alone. In 2023, additional confirmations regarding crop hosts included hydrangea (Hydrangeaceae) from New Jersey, U.S.A. [5]. However, the whereabouts of *S. dorsalis* after crop termination has been unclear. The refugia and food resources utilized by *S. dorsalis* during this period could shed further light on the population dynamics of this pest.

Therefore, the objective of this study was to identify alternative hosts of *S. dorsalis* being utilized after crop termination of strawberry and pepper. The presence of adults and larval stages of *S. dorsalis* on any wild plants over several summer months would indicate the expansion of its host range to include new reproductive hosts. Changes in population peaks would also indicate their migration patterns in and out of the alternate refuge areas.

2. Materials and methods

Scirtothrips dorsalis monitoring in woodland borders occurred in three commercial strawberry fields in Hillsborough County, FL, U.S.A. (28.0236825, −82.2036973) between June 2023 and July 2024. During both years, two conventional fields and one organic field were monitored for this study. Wild plants and trees were sampled once a month along the field borders based on the report that *S. dorsalis* tend to migrate into the outer edges of strawberry fields at the beginning of strawberry field season and dwell within 100 m from the field edge [6]. In 2023, fifteen plant species were collected (six weed species and nine tree species), and in 2024, seventeen species were collected (ten weed species and seven tree species). Plants were marked and numbered using marking tape to continue sampling from the same plants throughout the year. Five to six young leaves or trifoliates were collected per plant species per farm and stored in a sealed plastic bag. A pole saw (Hyper Tough, Libertyville, IL, U.S.A) was used to access young leaves from taller trees when appropriate. The samples were placed in a cooler and transported back to the entomology laboratory at the University of Florida, Gulf Coast Research and Education Center, FL, U.S.A. (27.7611241, −82.2302406), for processing. All samples were left to incubate for five to seven days after collection to allow for any *S. dorsalis* eggs to hatch. After incubation, the samples were washed with 70% ethanol solution and all *S. dorsalis* were counted and identified under a microscope with the use of taxonomic keys [7]. Absolute count of adult and larval *S. dorsalis* per plant species per field site was recorded. Plant species were confirmed using identification keys [8, 9]. The samples extracted from these collections were also submitted to the Florida Department of Agriculture and Consumer

Services (FDACS), Division of Plant Industry (DPI), Gainesville, FL, U.S.A., for species confirmation.

The absolute numbers of *S. dorsalis* adults and larvae are reported here as untransformed data.

3. Results

A sample of the thrips sent to FDACS-DPI was confirmed to contain *S. dorsalis* (Sample number 10062023–10412). During June to December 2023, a total of twenty wild plant species were sampled across the three commercial strawberry fields. Nine plant species, Spanish needles [*Bidens alba* (L.) DC.; Asteraceae], Darlington oak [*Quercus hemisphaerica* Bartram ex Willd.; Fagaceae], muscadine grape (*Vitis rotundifolia* Michx.; Vitaceae], Carolina Laurel cherry [*Prunus caroliniana* (Mill.) Aiton; Rosaceae], painted spurge [*Euphorbia heterophylla* L.; Euphorbiaceae], Peruvian water primrose [*Ludwigia peruviana* (L.) H. Hara; Onagraceae], ragweed [*Ambrosia psilostachya* L.; Asteraceae], sweetgum [*Liquidambar styraciflua* L.; Altingiaceae], and water oak [*Quercus nigra* L.; Fagaceae], had both adult and larval stages of *S. dorsalis* (**Table 1**). Two plant species, Laurel oak [*Quercus laurifolia* Michx.; Fagaceae], and sugarberry [*Celtis laevigata* Willdenow; Cannabaceae] had larval *S. dorsalis* only, whereas camphor-tree [*Cinnamomum camphora* (L.) J.Presl.; Lauraceae] had adults only. The eight wild plants that did not host *S. dorsalis* were beautyberry [*Callicarpa* spp. L.; Lamiaceae], bitter melon [*Momordica charantia* L., Cucurbitaceae], Caesar's weed [*Urena lobata* L.; Malvaceae], coffee senna [*Senna occidentalis* (L.) Link; Fabaceae], elderberry [*Sambucus canadensis* L.; Adoxaceae], southern live oak [*Quercus virginiana* Mill.; Fabaceae], southern red oak [*Quercus falcata* Michx.; Fagaceae], and sunn hemp [*Crotolaria juncea* L. Fabaceae].

Muscadine grapes and Carolina Laurel cherry had *S. dorsalis* adult and larvae during the entire non-crop season from June through October 2023. The occurrence of *S. dorsalis* on the other plant species was sporadic every month, with ragweed hosting 6 adults and 23 *S. dorsalis* larvae only in the month of July (**Table 1**). Only two of the three field sites had ragweed, and of those two, only one site had ragweed infested with *S. dorsalis*. Interestingly, during the month of November when *S. dorsalis* could no longer be detected on the wild plants, only four wild plants, Spanish needles, painted spurge, sugarberry, and Peruvian water primrose hosted *S. dorsalis*.

In 2024, monthly sampling continued from January through July covering seventeen plant species. Florida maple [*Acer saccharum* subsp. *floridanum* (Chapm.) Pax; Sapindaceae] was added to the sampling list. Southern live oak, southern red oak, Darlington oak, and sunn hemp were not sampled in 2024 due to lack or negligible occurrence of *S. dorsalis* on these species in 2023. Some field sites no longer had enough wild specimens of these plant species and were thus eliminated from the sampling list. The first record of *S. dorsalis* in 2024 was found on April 5th on Spanish needles, muscadine grape, sugarberry, and Laurel cherry.

In fact, on Peruvian water primrose, only *S. dorsalis* larva was recorded on this date, and both adults and larvae were recorded during rest of the sampling dates through July. An infestation on sweetgum was confirmed but not until May. In ragweed, both adult and larval *S. dorsalis* started appearing in June and July. Florida maple was found to host a few *S. dorsalis* adults and larvae only in July, not before. Overall, all the plants were found to host both adult and larval *S. dorsalis* populations.

Date	Plant	*S. dorsalis* adults	*S. dorsalis* larvae	Farm
June 5, 2023				
	Grape, Vitis rotundifolia Michx.	0	1	Farm 1
		0	6	Farm 2
	Laurel cherry, Prunus caroliniana (Mill.) Aiton	2	2	Farm 1
		1	13	Farm 2
	Sugarberry, Celtis laevigata Willdenow	0	5	Farm 2
	Darlington oak, Quercus hemisphaerica Bartram ex Willd.	0	3	Farm 3
July 7, 2023				
	Ragweed, Ambrosia psilostachya L.	6	23	Farm 1
	Grape, Vitis rotundifolia Michx.	2	4	Farm 1
		1	3	Farm 2
	Laurel cherry, Prunus caroliniana (Mill.) Aiton	9	12	Farm 1
		1	6	Farm 2
	Sweet gum, Liquidambar styraciflua L.	2	3	Farm 2
	Camphor tree, Cinnamomum camphora (L.) J.Presl.	1	0	Farm 2
	Water oak, Quercus nigra L.	1	2	Farm 2
	Darlington oak, Quercus hemisphaerica Bartram ex Willd.	1	1	Farm 3
Aug 3, 2023				
	Grape, Vitis rotundifolia Michx.	1	2	Farm 1
	Laurel cherry, Prunus caroliniana (Mill.) Aiton	1	3	Farm 1
		1	2	Farm 2
	Camphor tree, Cinnamomum camphora (L.) J.Presl.	1	0	Farm 2
	Water oak, Quercus nigra L.	1	1	Farm 2
Sep 6, 2023				
	Spanish needles, Bidens alba (L.)	0	2	Farm 1
		1	0	Farm 2
		0	1	Farm 3
	Grape, Vitis rotundifolia Michx.	1	11	Farm 1
	Laurel cherry, *Prunus caroliniana* (Mill.) Aiton	1	3	Farm 1
		1	1	Farm 2
	Laurel oak, Quercus laurifolia Michx.	0	1	Farm 2
Oct 10, 2023	Spanish needles, Bidens alba (L.)	0	2	Farm 1
		1	0	Farm 2
	Grape, Vitis rotundifolia Michx.	2	1	Farm 1
		0	3	Farm 2

Date	Plant	*S. dorsalis* adults	*S. dorsalis* larvae	Farm
	Laurel cherry, *Prunus caroliniana* (Mill.) Aiton	2	1	Farm 1
		2	5	Farm 2
Nov 2, 2023	Painted spurge, Euphorbia heterophylla L.	1	1	Farm 1
	Peruvian water primrose, Ludwigia peruviana (L.) H. Hara	2	3	Farm 1
	Grape, Vitis rotundifolia Michx.	0	3	Farm 2
	Spanish needles, Bidens alba (L.)	1	2	Farm 2
	Sugarberry, Celtis laevigata Willdenow	0	2	Farm 2
Apr 5, 2024	Spanish needles, Bidens alba (L.)	1	1	Farm 3
		1	1	Farm 2
	Peruvian water primrose, Ludwigia peruviana (L.) H. Hara	0	1	Farm 3
	Grape, Vitis rotundifolia Michx.	1	1	Farm 3
		0	1	Farm 2
	Sugarberry, Celtis laevigata Willdenow	1	1	Farm 2
	Laurel cherry, *Prunus caroliniana* (Mill.) Aiton	2	1	Farm 2
May 7, 2024	Spanish needles, Bidens alba (L.)	2	3	Farm 3
		1	2	Farm 2
	Peruvian water primrose, Ludwigia peruviana (L.) H. Hara	1	1	Farm 3
	Grape, Vitis rotundifolia Michx.	1	1	Farm 3
		1	1	Farm 2
	Sweetgum, Liquidambar styraciflua L.	1	1	Farm 2
	Sugarberry, Celtis laevigata Willdenow	1	1	Farm 2
	Laurel cherry, *Prunus caroliniana* (Mill.) Aiton	2	1	Farm 2
June 7, 2024	Spanish needles, Bidens alba (L.)	3	2	Farm 3
		4	1	Farm 2
	Peruvian water primrose, Ludwigia peruviana (L.) H. Hara	1	1	Farm 3
	Ragweed, Ambrosia psilostachya L.	2	1	Farm 3
	Grape, Vitis rotundifolia Michx.	2	1	Farm 2
	Sweetgum, Liquidambar styraciflua L.	2	2	Farm 2
	Sugarberry, Celtis laevigata Willdenow	2	1	Farm 2
	Laurel cherry, *Prunus caroliniana* (Mill.) Aiton	1	3	Farm 2
	Water oak, Quercus nigra L.	1	0	Farm 2
July 19, 2024	Spanish needles, Bidens alba (L.)	2	2	Farm 3
		3	2	Farm 2
	Peruvian water primrose, Ludwigia peruviana (L.) H. Hara	3	2	Farm 3

DOI: http://dx.doi.org/10.5772/intechopen.1010077

Date	Plant	*S. dorsalis* adults	*S. dorsalis* larvae	Farm
	Ragweed, Ambrosia psilostachya L.	1	1	Farm 3
	Grape, Vitis rotundifolia Michx.	2	3	Farm 3
		2	2	Farm 2
	Florida maple, Acer saccharum subsp. *floridanum* (Chapm.) Pax	1	1	Farm 3
	Sweetgum, Liquidambar styraciflua L.	1	2	Farm 2
	Sugarberry, Celtis laevigata Willdenow	2	2	Farm 2
	Laurel cherry, *Prunus caroliniana* (Mill.) Aiton	2	2	Farm 2
	Water oak, Quercus nigra L.	1	1	Farm 2

Table 1.
Number of S. dorsalis *adult and larvae on each sampling date and plant species in June 2023–July 2024. Sampling dates across Farms 1, 2, and 3 in Hillsborough County, Florida, U.S.A are June 5, July 7, August 3, September 6, October 10, November 2, December 4 in 2023 and January 3, February 5, March 5, April 5, May 7, June 7, and July 19 in 2024. Only plant species with confirmed* S. dorsalis *adult and larval populations are reported in this table. Only sampling dates when* S. dorsalis *were confirmed are reported here.*

The plant species not hosting *S. dorsalis* during January to July 2024 sampling event were beautyberry, bitter melon, camphor, Caesar's weed, coffee senna, elderberry, Laurel oak, and painted spurge.

4. Discussion

This field study shows that the wild plants in woodland borders around strawberry fields of Florida start getting infested with *S. dorsalis* around first week of April and the infestation within the woodland may persist till first week of November. This is in synch with the strawberry crop season as the strawberry crop in FL, U.S.A., spans between September to May [10]. As the strawberry crop season comes to an end, *S. dorsalis* start migrating back into the woodland areas around these fields to colonize certain plant species in April. In the following strawberry crop season, as the bulk of strawberry transplanting and crop development begins in September to early October, *S. dorsalis* start migrating back into the strawberry crop. This points to the consistent proximity of *S. dorsalis* to the annual strawberry crop in FL, and the ability of this invasive species to capitalize on natural resources when its preferred host is absent.

Previous studies have reported that *S. dorsalis* prefer host plants with lower height versus taller crops and have a weak dispersal capability. According to one study, during six weeks, *S. dorsalis* moved a maximum of 12 m from their reservoir population [11]. In strawberry fields, *S. dorsalis* migrated to a maximum of 100 m inside the fields over four months from field border and its population hotspots remain aggregated mostly within the 50 m distance from the field border [4]. In open-field strawberry, *S. dorsalis* clearly showed a preference for the field edge over field interior areas [4]. Kumar et al. [11] also reported that flight activity of *S. dorsalis* was found to be the highest between 10:00 AM and 4:00 PM where the peak solar radiation was recorded at ~337–653 w/sq. m. With limited flight activity, *S. dorsalis* prefers to inhabit the field edge ecosystem where they can easily migrate in and out of crops when crop health and phenology is conducive for their rapid growth.

Wild plant species such as muscadine grape, Laurel cherry, sugarberry, Spanish needles, Peruvian water primrose, ragweed, sweetgum, and water oak showed consistent infestation with *S. dorsalis* during both summer non-crop seasons in this study. The plant species not hosting *S. dorsalis* during both non-crop seasons in 2023 and 2024 were beautyberry, bitter melon, camphor, Caesar's weed, coffee senna, elderberry, and Laurel oak. The painted spurge weed was unique because *S. dorsalis* infestation was confirmed from it in 2023 but not in 2024. Florida maple was only sampled in 2024 summer and *S. dorsalis* was found reproducing on this plant. The information regarding successfully used wild host plants has important implications on management tools for *S. dorsalis*. These plant species can be used for a targeted augmentative biological control program where biological control agents can be specifically deployed on these plants to facilitate extended interaction with *S. dorsalis* and avoid non-target impacts. This could also lower the cost of pesticide application for growers during crop season. These alternative wild hosts of *S. dorsalis* should be considered as ecosystem resources, where natural enemies of *S. dorsalis* can freely interact with a persistent pest population without experiencing lethal/sub-lethal non-target effects of conventional insecticides.

The wide host range of *S. dorsalis* is well documented among crop hosts [5, 12]. This study contributes to our understanding of host range expansion capabilities of *S. dorsalis* with regards to wild plants or non-crop hosts within its new invasive region of the U.S.A. This information is also relevant to the other invasive ranges of *S. dorsalis* in the world as these plant families are commonly found across the globe. Plant families previously reported within the U.S.A host range of *S. dorsalis* that also occurred in this field study are Asteraceae (false daisy, *Eclipta prostrata* (L.); marigold little orange, *Tagetes erecta* L.; zinnia Zahara yellow, *Zinnia marylandica*; common yarrow, *Achillea millefolium*; osteospermum daisy bushes, *Osteospermum ecklonis*; Spanish needle; purple coneflower, *Echinacea purpurea*), Malvaceae (cotton, *Gossypium hirsutum* L.; hibiscus, *Hibiscus rosa-sinensis*), Fabaceae (thirteen species including peanut, groundnut, *Arachis hypogaea* L.; soybean, *Glycine max* (L.) Merr.), Rosaceae (rose, *Rosa* spp.; peach *Prunus persica* (L.) Batsch; *Pyrus* spp.; *Rubus* spp.), and Sapindaceae (litchi, lychee, *Litchi chinensis* Sonn.) [5, 12–14].

Figure 3.
Adult minute pirate bug, Orius insidiosus *Say feeding on an* S. dorsalis *larva. Image credit: Lovely Adhikary, UF/IFAS GCREC.*

Figure 4.
Adult big-eyed bug, Geocoris *spp. Image credit: Chastity Perry, UF/IFAS GCREC.*

However, some new plant families being reported from U.S.A for the first time in this study as hosts of *S. dorsalis* include Fagaceae, Onagraceae, Altingiaceae, Cannabaceae, and Lauraceae. This information is useful because a survey of natural enemies can now be targeted, and conservation of natural enemies can be planned based on the occurrence of these plant species in woodlands. These natural enemies include *Orius* spp. (Hemiptera: Anthocoridae) (**Figure 3**) and *Geocoris* spp. (Hemiptera: Geocoridae) (**Figure 4**).

Integrated pest management of invasive species needs to involve non-crop seasons and woodland border areas, so that farms can be more resilient, relying less on chemical control tactics alone. Knowledge of plant hosts being utilized by invasive species throughout the year can improve IPM of pests during the crop season. The baseline information provided in this study can benefit IPM for *S. dorsalis* within the U.S.A., and its new global range.

5. Conclusion

This book chapter reports the plant hosts discovered to be serving as the reproductive hosts of the invasive thrips species, *Scirtothrips dorsalis* Hood in Florida, U.S.A. The report is based on a two-year field study in woodland border areas of commercial strawberry fields, where *S. dorsalis* have established themselves as an economically devastating pest of the strawberry crop. The findings in this study have far-reaching implications for integrated pest management, especially augmentative and conservation biological control efforts to manage this invasive pest within U.S. and all global regions falling within the invasive range of *S. dorsalis*.

Acknowledgements

The authors thank Allan Busuulwa for assistance with field sampling. Also, authors are thankful to Felipe Soto-Adames and Alex de la Paz of Florida Department of Agriculture and Consumer Services Department of Plant Industry for thrips and plant species identification. We also thank the Florida Strawberry Research Education Foundation Inc., Florida Foundation Seed Producers, Inc., and USDA National Institute of Food and Agriculture [Hatch] project [FLA-GCR-005888] for funding this research.

Conflict of interest

The authors declare no conflict of interest.

Author details

Sriyanka Lahiri* and Chastity Perry
Gulf Coast Research and Education Center, University of Florida, Florida, U.S.A.

*Address all correspondence to: lahiris@ufl.edu

References

[1] Lahiri S, Orr D, Sorenson C, Cardoza Y. Overwintering refuge sites for *Megacopta cribraria* (Hemiptera: Plataspidae). Journal of Entomological Science. 2015;**50**:69-73. DOI: 10.18474/0749-8004-50.1.69

[2] Tillman G, Toews M, Blaauw B, Sial A, Cottrell T, Talamas E, et al. Parasitism and predation of sentinel eggs of the invasive brown marmorated stink bug, *Halyomorpha halys* (Stål) (Hemiptera: Pentatomidae), in the southeastern US. Biological Control. 2020;**145**:104247. DOI: 10.1016/j.biocontrol.2020.104247

[3] USDA-NASS. Southern Region News Release, Noncitrus Fruits and Nuts. Washington, D. C.: U. S. Department of Agriculture, National Agricultural Statistics Service; 2023. Available from: https://www.nass.usda.gov/Statistics_by_State/Regional_Office/Southern/includes/Publications/Crop_Releases/Fruit_Production/NoncitrusFruitNut2023.pdf [Accessed: January 30, 2025]

[4] Kaur G, Stelinski LL, Martini X, Boyd N, Lahiri S. Reduced insecticide susceptibility among populations of *Scirtothrips dorsalis* Hood (Thysanoptera: Thripidae) in strawberry production. Journal of Applied Entomology. 2023;**147**:271-278. DOI: 10.1111/jen.13108

[5] Kumar V, Xiao Y, Borden MA, Ahmed MZ, McKenzie CL, Osborne LS. Distribution of *Scirtothrips dorsalis* (Thysanoptera: Thripidae) cryptic species complex in the United States and reproductive host assessment of its dominant member. Journal of Economic Entomology. 2023;**116**:1715-1726. DOI: 10.1093/jee/toad138

[6] Kaur G, Stelinski LL, Martini X, Boyd N, Mallinger R, Lahiri S. Spatial and temporal distribution of *Scirtothrips dorsalis* (Thysanoptera: Thripidae) and their natural enemies in Florida strawberry fields. Journal of Economic Entomology. 2024;**117**:1418-1429. DOI: 10.1093/jee/toae111

[7] Cluever JD, Smith HA. A photo-based key of thrips (Thysanoptera) associated with horticultural crops in Florida. Florida Entomologist. 2017;**100**:454-467. DOI: 10.1653/024.100.0208

[8] Murphy TR, Colvin DL, Dickens R, Everest JW, McCarty LB. Weeds of southern turfgrasses. In: The University of Florida, Cooperative Extension Service. Gainesville, FL, USA: Institute of Food and Agricultural Services; 2004. p. 208. Available from: https://original-ufdc.uflib.ufl.edu/AA00091867/00001

[9] Butler L, Davis T, Hogue P, Russakis N, Sellers B. Pasture weeds of South Florida. In: The University of Florida, Cooperative Extension Service. 3rd ed. Gainesville, FL, USA: Institute of Food and Agricultural Services; 2020. p. 42. Available from:https://rcrec-ona.ifas.ufl.edu/media/rcrec-onaifasufledu/pdf/Pasture-Weeds-of-South-Florida-3rd-ed-2020.pdf

[10] Whitaker VM, Peres N, Lahiri S, Brown SP, Chandler CK. Growing strawberries in the Florida home garden. In: The University of Florida, Cooperative Extension Service. Gainesville, FL, USA: Institute of Food and Agricultural Services; 2021. p. HS1154. DOI: 10.32473/edis-hs403-2021

[11] Kumar V, Kakkar G, Seal DR, McKenzie CL, Colee J, Osborne LS.

Temporal and spatial distribution of an invasive thrips species *Scirtothrips dorsalis* (Thysanoptera: Thripidae). Crop Protection. 2014;**55**:80-90. DOI: 10.1016/j.cropro.2013.10.015

[12] Kumar V, Seal DR, Kakkar G, McKenzie CL, Osborne LS. New tropical fruit hosts of *Scirtothrips dorsalis* (Thysanoptera: Thripidae) and its relative abundance on them in South Florida. Florida Entomologist. 2012;**95**:205-207. DOI: 10.1653/024.095.0134

[13] Amin BW. Techniques for handling thrips as vectors of tomato spotted wilt virus and yellow spot virus of groundnut, *Arachis hypogea* L. L. Occasional Paper. Groundnut Entomology ICRISAT. 1980;**80**:1-20

[14] Kumar V, Seal DR, Kakkar G. Chilli thrips – *Scirtothrips dorsalis* Hood. Featured Creatures. Gainesville, FL, USA: University of Florida, Institute of Food and Agricultural Services; 2009. EENY-463. Available from: https://entnemdept.ufl.edu/creatures/orn/thrips/chilli_thrips.htm

Chapter 7

The Role of Metals in Biological Control of Insects

Inam Ullah, Hafza Wajeeha Ijaz, Yudie Ma and Minglin Lang

Abstract

Metals are essential to insect physiology and play a crucial role in biological control strategies for managing pest populations. This chapter explores the dual roles of metals as vital micronutrients and disruptors of insect homeostasis, focusing on their uptake, regulation, and detoxification mechanisms in pest and beneficial insects. We examine the influence of metals on insect development, reproduction, and immunity, with an emphasis on their role in metalloprotein enzymes. Additionally, the chapter delves into metal-based pest-control strategies, including metal-containing pesticides, nanoparticles, and their integration with biological control methods. Emphasis is placed on the synergistic effects of metals with natural biocontrol agents and their role in integrated pest management. Challenges, such as environmental accumulation, resistance development, and non-target toxicity, are addressed alongside sustainable approaches and emerging biotechnologies like CRISPR (Clustered Regularly Interspaced Short Palindromic Repeats) and RNAi (RNA interference). By bridging the gap between metal regulation and pest control, this chapter provides novel insights into leveraging metals for effective, eco-friendly pest management.

Keywords: insect physiology, nanoparticles, homeostasis, agriculture, predatory insects

1. Introduction

The loss of crops to pests and diseases is the primary worry of farmers, irrespective of whatever production strategy they use. A pest is any undesired insect that harms and competes with crops. Insects, pests, and weeds annually destroy 40% of all food crops [1, 2]. Various traditional pesticides, including carbamates, organophosphates, and organochlorines, were created [3]. The pest management industry is continuously looking for new ways to manage pests. Predators, parasites, and parasitoids, collectively called "natural enemies," which attack and kill pest populations frequently in nature are also exploited to enhance their pest predation potential, and some plants develop resistance to some pests [4–6]. Pests limit plant density, restrict plant growth, and produce less than what is needed. Several strategies are adopted for pest management depending on the circumstances, the kind of pests, and the degree of intervention needed. Among those cultural pest management techniques are crop rotation, appropriate irrigation, sanitation, pruning, and companion planting [7].

Metals are used to intervene in chemical or biological processes. Both essential and non-essential metals have an intriguing and frequently overlooked role in the biological regulation of insects. Pesticide formulations frequently contain metals, and some metallic compounds are naturally insecticidal [8, 9]. These substances can interfere with the physiological functions, behavior, and metabolism of insects. Metals also have an impact on several beneficial organisms utilized in biological pest control, either directly or indirectly through their operating environment. The presence or concentration of metals in the environment can affect how well natural enemies, such as parasitoids, predators, and pathogens, control pests. For instance, these helpful organisms' behavior or ability to survive can be changed by metal contamination in soil or water [10, 11].

It has been discovered that certain metals interact with metabolic pathways to affect insect development and reproduction. Insect populations may be lowered as a result of this interaction, which may either slow growth or speed up the life cycle [9]. Some studies have suggested that higher concentrations of metals can affect insect molting and metamorphosis, leading to developmental failure [12]. Some metal-based interventions can disturb calcium homeostasis, making insects more vulnerable to dehydration or predation. Certain metals can change the behavior of insects, which could result in fewer of them or make it simpler for predators or traps to control them [13]. Some metals affect the attraction or repulsion of insects. For example, zinc has been shown to interfere with some insect species' sensory receptors, making it more difficult for the insects to find food or potential mates. Similarly, some metals can impair an insect's capacity to produce or sense pheromones, which are necessary for communication and mating [14, 15]. Metals used in pest control must be used carefully to reduce hazards to the environment and human health. Investigation into the safe and efficient use of metal-based techniques may lead to integrated pest management and environmentally friendly replacements for the use of conventional chemical pesticides.

2. Metal ion transporters and homeostasis

Integral membrane proteins that transport metal ions across biological membranes are known as metal ion transporters. These transporters are critically important for proper cellular function and overall physiological balance in insects. They are responsible for the uptake, distribution, and excretion of essential metal ions, including iron (Fe), copper (Cu), zinc (Zn), and manganese (Mn), and for the regulation of potentially toxic metals, such as cadmium (Cd) and lead (Pb) [16]. Insect physiology depends on metal ion homeostasis, a fine balance of essential metal ions in cells and tissues. Many biological functions depend on this process, which is highly regulated and must be maintained at optimal ion concentrations to avoid toxicity [9]. Understanding the structure-function relationship of these metal transporters and homeostatic mechanisms is not only important for understanding insect and microbial pest physiology, but also has potential applications in pest management and environmental monitoring, as shown in **Figure 1**. Disrupting metal ion homeostasis could be a target for novel pesticides.

2.1 Metal uptake

Insects normally absorb metals after eating through their midgut. The intestinal epithelium allows metal ions to pass into the hemolymph (insect blood). Strict control is made on this uptake to make sure that only essential metals, like Fe, Cu, or Zn, are absorbed at the proper levels and that potentially dangerous metals, like Cd, are

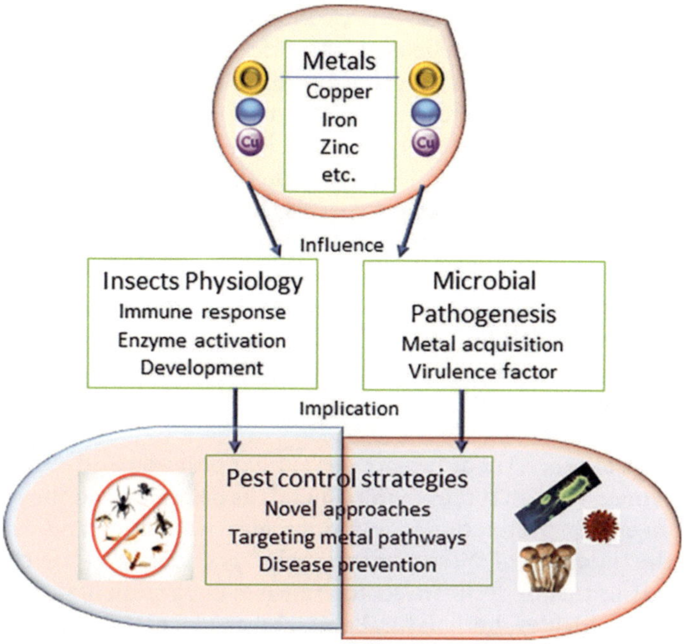

Figure 1.
Influence of metals on insect physiology and pest control. Insect physiology (immune response, enzyme activation) and microbial pathogenesis (metal acquisition, virulence) are influenced by metals, such as Cu, Fe, Zn, etc. These interactions offer avenues to novel pest-control strategies aimed at disease prevention and pest management through the manipulation of metal pathways.

reduced. Some ion channels and transporters on the gut epithelium regulate metal absorption, including zinc transporter (ZnT) or divalent metal transporter 1 (DMT1). These transporters make selective uptake possible because they can identify specific metals. Some insects have gut-chelating substances that bind to metals and prevent excessive absorption. For example, phytochelatins, small peptides secreted by the digestive system, can bind heavy metals to prevent the absorption of toxic metals [17, 18].

2.2 Transport and distribution of metals

The hemolymph distributes metals throughout the body after they pass through the digestive tract. Often, the movement of metals within the hemolymph is mediated by metal-binding proteins. These proteins can bind metals and prevent the free ionic form of metals from interacting with biological structures and causing damage. These tiny, cysteine-rich proteins bind and store metals, such as Zn, Cu, and Cd [16]. Metallothionein is necessary for buffering the amounts of metals in tissues, detoxifying excess metals, and storing them. Insects use certain transport proteins to manage the distribution of vital metals, such as Cu and Zn. ZnT and Cu transporting adenosine triphosphatases (ATPases) transport these metals into cells, where they are either stored for future use or incorporated into metalloenzymes [19].

2.3 Metal storage and detoxification

Insects tend to manage excess metals by sequestering them inside specialized organs, such as the fat body, malpighian tubules, and midgut cells, isolating the metals in granules or vesicles to prevent toxic interactions with biological components. Metallothionein and phytochelatins comprise peptide chains that store metals such as Cu and Cd, lowering their free concentration and thereby preventing cellular damage. Metals can also be excreted through the malpighian tubules to prevent accumulation, especially for toxic metals like Cd [20]. Insects employ antioxidant enzymes, such as catalase, glutathione-peroxidase (GPx), and superoxide dismutase (SOD), to neutralize the reactive oxygen species (ROS) synthesized during oxidative stress brought up by the metals. In some species, an increase in transcript levels of copper-zinc superoxide dismutase (Cu/Zn-SOD) relates to Cu excess, whereas glutathione -S-transferases (GSTs) couple metals to glutathione for removal in a detoxification pathway [21, 22].

2.4 Metal-dependent enzymes

The physiological processes of insects and similar pests depend on metal-dependent enzymes. Such enzymes require metals as cofactors and they operate on everything from energy production to immunological protection and detoxification [9]. Knowing the functions of these enzymes helps us understand insect biology and the mechanisms that underlie insect tolerance to environmental stresses, such as heavy metals and pesticides, both natural and man-made. Metal-dependent enzymes are also among the components contributing to insecticide resistance. Many insect species are now capable of expressing iron-dependent cytochrome P450 enzymes, enabling them to detoxify such harmful substances as insecticides [23]. Cu/Zn-SOD also helps in reducing oxidative stress brought on by pesticides [22]. When these enzymes are overexpressed, resistance may advance, making chemical pest population management more challenging. Additionally, one significant way that certain pests, particularly those found in metal-polluted areas, have developed to survive in contaminated settings is the capacity of metal-dependent enzymes, such as metallothionein, to sequester and detoxify heavy metals and xenobiotics [24].

Metal ion transporters and homeostatic mechanisms are vital for insect physiology, ensuring proper cellular function, detoxification, and survival under various environmental stresses. A deeper understanding of these processes offers valuable insights into insect biology and potential strategies for pest control and environmental sustainability.

3. Metals as tools for biological control

As we know, metals play an integral role in the physiology of insects; however, excessive amounts of metals can disrupt critical physiological processes of insects, including respiration, enzyme activity, and reproduction [25, 26]. In recent years, there has been growing interest in using metals and metal-based compounds as alternative pest-control agents. Traditionally, synthetic chemical pesticides have dominated agricultural pest management, offering broad-spectrum efficacy but often posing risks to human health, beneficial organisms, and the environment. In contrast, metals present a novel approach that influences their essential biochemical roles and toxic potential in low doses. Despite that, metals should be used with great care as pesticides [27, 28]. Over time, metals can accumulate in soil and water systems resulting in environmental problems. As a result, there is much need for the development of biodegradable metal-based formulations or controlled release systems to maintain this approach sustainable.

3.1 Metal-based pesticides

The concept of using metals as pesticides is rooted in their ability to interfere with insect physiology in many different ways. Metals like Cu, Zn, and silver (Ag) have long been known for their antimicrobial properties, and their use in agriculture as fungicides and bactericides is well established. Cu-based compounds, for instance, have been used as early as the nineteenth century to manage fungal diseases [29, 30]. These metals disrupt essential cellular functions, such as membrane integrity and enzyme activity, effectively controlling a wide range of agricultural pests with lower ecological impact compared to many synthetic pesticides. Metals can induce oxidative stress by generating ROS within pest cells, damaging cellular components like DNA, proteins, and lipids. Fe and Mn, essential in small amounts, can be lethal at higher concentrations, leading to disruptions in insect growth and reproduction, as shown in **Figure 2**. This targeted impact on metabolic processes makes metals a valuable tool for integrated pest management (IPM) strategies that aim to reduce chemical reliance and enhance environmental sustainability [31, 32]. Certain metals and metal compounds have been used as pesticides due to their toxic effects on pests, pathogens, and weeds. Here are some commonly used metals with pesticidal properties:

3.1.1 Copper

Copper (Cu) compounds, such as copper sulfate, copper hydroxide, and Bordeaux mixture (a blend of copper sulfate and lime), are extensively used as fungicides and bactericides in agriculture, especially in vineyards and orchards. Cu ions interact with proteins and disrupt enzyme function in microbial cells, leading to cell damage and death. These compounds prevent fungal spores from germinating and protect crops

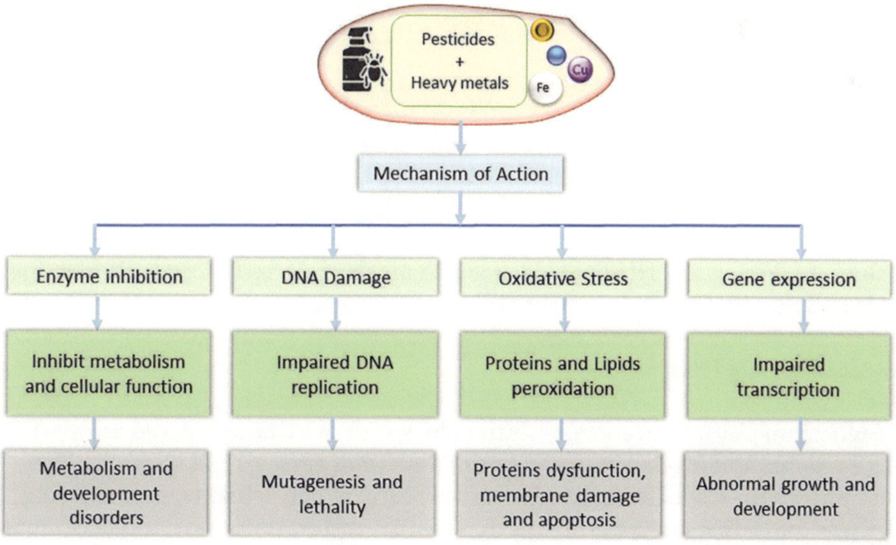

Figure 2.
Mechanisms of pesticide and heavy metal toxicity in insects. Pesticides and heavy metals disrupt insect physiology through enzyme inhibition, DNA damage, oxidative stress, and impaired gene expression, leading to metabolic disorders, mutagenesis, apoptosis, and abnormal development.

from diseases like downy mildew and leaf spots. However, Cu can accumulate in soil over time, so its use must be carefully regulated to avoid toxicity to plants and beneficial organisms [33].

3.1.2 Silver

Ag, especially in nanoparticle form, has potent antimicrobial and insecticidal effects. Ag nanoparticles (AgNPs) can penetrate cell membranes, generating ROS that damage cellular components like DNA, proteins, and lipids. This disrupts vital functions in pathogens and insect pests. Ag-based pesticides are effective against bacteria, fungi, and some insects, though they are used sparingly due to concerns over potential environmental toxicity and bioaccumulation [34].

3.1.3 Zinc

Zinc phosphide is primarily used as a rodenticide. When consumed, it reacts with stomach acid in rodents to produce phosphine gas, which is highly toxic and leads to respiratory failure. Zn phosphide is fast-acting and effective in rodent control, especially in settings where rodent populations pose significant risks to crops and stored grains. To mitigate risks to non-target species, Zn phosphide is generally used in bait formulations with strong attractants for rodents [35].

3.1.4 Boron

Boron compounds, particularly boric acid and borax, are used as insecticides against pests like ants, cockroaches, and termites. Boron disrupts digestive enzymes in insects, leading to malnutrition and death over time. It also damages their exoskeletons, causing dehydration. Boric acid is applied in powder or gel form in pest-prone areas and is especially valued for household pest control due to its low toxicity to humans and pets when used properly [36, 37].

3.1.5 Iron

Iron phosphate is commonly used as a molluscicide to control slugs and snails, which are serious pests in gardens and farms. Fe phosphate interferes with the digestive process of these pests, rendering them unable to consume food and finally causing death. Iron phosphate is often coupled with other substances that attract snails and slugs, making it an efficient bait. It is also much preferred for organic farming due to its low effects on the environment and minimal toxicity to non-target species [38].

3.1.6 Aluminum and magnesium

Magnesium phosphide and aluminum phosphide are fumigants and are used to control rodents and insects in stored grains and other agricultural harvests. They release phosphine gas when reacted with moisture, which is highly toxic to living organisms, including pests by obstructing cellular respiration. It works very fast and penetrates deep into the stored materials, but requires trained personnel due to its danger and exposure risks [39, 40].

These metal-based pesticides offer different mechanisms and can be highly effective in pest control. However, one needs to be very careful in using metals to avoid inadvertent accumulation or toxicity in non-target organisms. High concentrations of metals could harm the beneficial organisms while also disrupting soil health or leaching into water systems. Consequently, research and development efforts have been devoted to the creation of controlled or biodegradable metal-based supplements for minimal environmental impact while ensuring the availability of metals for beneficial organisms.

3.2 Metal nanoparticles as pesticides

Metal nanoparticles have attracted the attention of a lot of researchers as novel agents in pest control, owing to their unique characteristics, which make them more effective than their traditional metal-based counterparts. These agents include Ag, Cu, Fe, and Zn oxide nanoparticles, which are all very reactive, possess high surface-to-volume ratios, and can directly interact with cellular structures in insects. This makes them efficient at very low concentrations with significantly reduced unwanted effects on non-target environment [41].

The best among the types of nanoparticles that have been established as insecticides is Ag nanoparticles (AgNPs). The mechanism of action involves the production of ROS, membrane damage, and alteration of respiration, which cause different cellular dysfunctions. Such diverse actions reduce and make AgNPs highly useful against various insect pests and microbial pathogens, which are quite beneficial in crop protection [42]. Similarly, copper nanoparticles (CuNPs) are also very competent because they intrude insect molting and growth processes. They disrupt very crucial metabolic enzymes and cellular structures that could lead to insect death. In horticultural applications, CuNPs have been effective because they reduce fungicide and insecticide applications [43]. However, Fe oxide nanoparticles, Fe_2O_3, promise much in pest management because of their ability to induce oxidative stress and metal dysregulation. They can also generate ROS under suitable conditions, which further injure the insect's cellular structures and immune systems [44]. Another noteworthy addition concerning pest control is Zn oxide and titanium dioxide (TiO_2) nanoparticles. Added to the effects of their ultraviolet (UV)-induced toxicity, they also present alternate options for insect control, especially in laboratory studies that point toward their possible application in the field [45].

Nanoparticle technology is under investigation as a possible solution for sustainable pest control. New metal and alloy nanoparticle formulations are being prepared for effective pest control combined with the least environmental impact. Nanoparticles have several advantages, such as the reduced use of pesticides and lower risk toward beneficial insects. Still, further research is needed to assess the long-term ecological effects of nanoparticles and the optimization of their use in agriculture.

3.3 Enhancing efficacy of natural predators and parasites

Biological control agents are the key actors in pest management in sustainable agriculture. Among these, natural predators and parasites are considered the most promising for pest control. Predatory insects, parasitoids, and entomopathogenic fungi and bacteria are included as biological control agents in the IPM systems. However, their potency in pest control relies also on the environment and the availability of essential nutrients [46, 47]. Recent research has underscored the possible role of metals as enhancers of these biological control agents, adding a new dimension to pest-control strategies that are both effective and environmentally friendly. Although indirectly, metals can help in pest control by providing these vital host metals for growing beneficial insects and microorganisms. Trace metals are required for the metabolism and pathogenicity of some bacteria, fungi, and other biological control agents [48–50]. Then, by providing these metals in specific ways, we can encourage the beneficial organisms that naturally eat, or outcompete, pests, allowing for a balanced ecosystem approach to pest management. For example, some biocontrol bacteria and fungi must rely on the presence of trace metals to process their metabolism, and supplying these can enhance their biocontrol activity.

Some of the metals are important to the growth and survival of beneficial organisms used in biological pest control. These metals play dynamic roles in various biological processes, such as immunity and metabolism [51]. By strategically adding these metals, it is possible to improve the performance of these advantageous organisms. This can lead to improved pest control by increasing their ability to locate, attack, and suppress pests. Additionally, metals can indirectly support biological control by creating environments that favor natural enemies over pests and by enhancing the resilience of beneficial microbes and fungi. This can make them more effective in combating and inhibiting pest populations.

4. Other pest-control strategies

In addition to using metal ions for pest control, there are many other methods for controlling agricultural and forestry pests, as follows;

4.1 Physical and chemical control strategies

Chemical control uses pesticides to kill pests by applying them to crops, soil, or water. Although effective in the short term, long-term reliance harms ecosystems and raises food safety concerns. Excessive pesticide use in citrus farming increases costs and leaves harmful residues, impacting crop quality and consumer health. To minimize these issues, chemical methods should be cautiously applied and integrated with other pest-control strategies [52]. Chemical pesticides are classified by target pests: insecticides control insects by affecting their nervous systems; fungicides prevent fungal diseases by inhibiting growth; and herbicides control weeds by disrupting plant growth processes, as shown in **Figure 3**.

Physical pest control is an eco-friendly method involving mechanical trapping, seed selection, lasers, and other technologies. These methods effectively reduce chemical pesticide use and protect the environment. Mechanical trapping, such as sticky boards and trap nets, leverages pest behaviors to capture them efficiently. According to Upama Adhikari (2022), this approach is simple, environmentally safe, and a viable alternative to chemical methods; however, it is not very efficient. High-temperature exposure and lime water dipping target pest eggs in seeds [53]. Control over temperature, especially in greenhouses and storage, severely limits pest reproduction and survival. New technologies, like lasers and infrared surveillance, are also emerging that will prove very effective in the near future [54].

4.2 Agricultural control strategy

Agricultural pest control leverages ecological adjustments to inhibit pest survival and reproduction, offering cost-effective and sustainable practices. Key

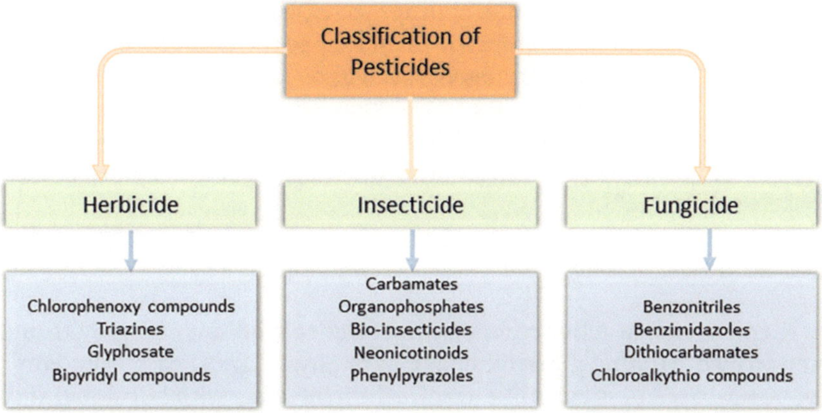

Figure 3.
This diagram shows the three main types of pesticides: herbicides, insecticides, and fungicides, each with specific chemical classes. Herbicides include chlorophenoxy compounds, triazines, glyphosate, and bipyridyl compounds. Insecticides are classified into carbamates, organophosphates, bio-insecticides, neonicotinoids, and phenylpyrazoles. Fungicides include benzonitriles, benzimidazoles, dithiocarbamates, and chloroalkylthiol compounds.

methods involve improving cultivation or weed control methods, such as crop rotation, plowing, drainage, and fertilizer, to regulate planting periods. IPM is the use of models to reduce dependence on chemical pesticides and enhance crop resilience in a sustainable framework that leads to increased production. Upgraded planting methods, such as the intercropping of U.S. Midwest corn with soybeans, enhance soil quality and decrease pests, and intercropping of vegetables and herbs has been promising in Asia and Europe [55, 56]. Furthermore, the planting of weed belts provides habitats for natural enemies, such as parasitic wasps and predatory insects, improving pest-control effectiveness and enhancing ecological balance and stability [57].

4.3 Biological control strategies

Biological control uses natural enemies, pathogens, or parasites to manage pests, offering an environmentally friendly alternative to chemical pesticides. Avoidant plants, biocontrol agents, natural enemies, plant extracts, and emerging biotechnologies, such as biopesticides and genetic control, are common biological control tools.

4.3.1 Avoidant plants

Avoidant plants repel pests by releasing chemicals like limonene, which inhibits pest activity and attracts natural enemies [10]. Limonene's dual role in repelling pests and enhancing ecosystem pest control by attracting natural enemies like parasitic wasps is widely used in agriculture [58].

4.3.2 Biocontrol agents

Natural enemies, such as predatory insects, parasitic wasps, and predatory mites, play key roles in pest management. Ladybird beetles and lacewings reduce aphid populations, while parasitic wasps like red-eyed wasps control pests, such as rice leaf borers and corn borers, effectively in agroecosystems [59]. Enhancing crop diversity through trap crops or aromatic plants supports natural enemy populations, improving pest control. Pathogens, such as baculoviruses, bacteria (e.g., *Bacillus thuringiensis* (Bt)), fungi (e.g., *Beauveria bassiana*, *Metarhizium* spp.), and nematodes, parasitize and kill pests, reducing reliance on chemical pesticides. For instance, Bt toxin disrupts pest digestion, while fungi secrete enzymes that degrade pest exoskeletons [60, 61].

4.3.3 Plant extracts

Plant extracts, rich in active compounds, effectively inhibit pest growth and reproduction by disrupting digestive enzymes or physiological pathways. For example, neem-based pesticides like azadirachtin interfere with hormonal and developmental processes in pests, reducing reliance on chemical pesticides [62]. Essential oils exhibit insecticidal properties by disrupting pest nervous and endocrine systems. Recent plant-derived formulations like S-enemas mimic insect hormones to control pest populations, showing promising results in pest management [63].

4.3.4 Biotechnology applications

Biotechnology offers innovative solutions for sustainable pest management. Natural or engineered biopesticides, such as microbial-based repellents and plant-derived compounds, reduce reliance on chemical pesticides. Biofumigation, utilizing gases from microbial metabolites, effectively targets soil pests by poisoning or disrupting cellular structures [64].

Genetic control technologies provide advanced methods for pest suppression. Transgenic crops, such as Bt maize and cotton, resist pests by producing insecticidal proteins. CRISPR-Cas9 (CRISPR-associated protein 9) gene editing enables precise genetic modifications, like altering zebra butterfly gut microbes to hinder development [65]. The Sterile Insect Technique, which releases sterile males to reduce pest reproduction, has proven effective in Mediterranean fruit fly control [66]. Gene-drive technology rapidly spreads pest-suppressing traits, such as in mosquito control [67]. RNA interference (RNAi) uses double-stranded RNA (dsRNA) to silence genes, effectively inhibiting pest growth with minimal environmental impact [68].

Integrating Internet of Things (IoT), drones, Artificial Intelligence (AI), and nanotechnology, precision tools enable real-time pest monitoring and targeted interventions. Unmanned aerial vehicles (UAVs) optimize pesticide spraying, while AI-powered systems identify pests effectively [69, 70]. Innovative plant-based solutions and biotechnological advancements significantly improve pest-control efficiency while minimizing environmental impact and reducing chemical pesticide dependency.

5. Methods for metal production and evaluation

5.1 Sourcing metals for pesticidal applications

The metals used in pesticidal applications are obtained from a variety of sources, depending on their availability, cost-effectiveness, and suitability for agricultural purposes. Common sources include:

5.1.1 Mining and refining

The ores yielding various metals, for example, copper, zinc, and arsenic, are mined and refined to produce pure metals. For instance, copper, widely used in pesticides like copper sulfate, is mined and processed into its pesticidal form through chemical treatments. Most zinc compounds are also products of ores mined, after which they are subsequently transformed into final products, such as zinc oxide, which possibly has applications as a fungistatic and bacteriostatic agent [71].

5.1.2 Recycling

During the time of increasing environmental concerns, recycling metals turned to become one important source. Recycled copper, zinc, and other metals can be repurposed for agricultural uses, reducing dependence on virgin materials and lowering environmental footprints [72].

5.1.3 Industrial by-products

The smelting of gold, silver, or copper results in the generation of by-products, such as arsenic and cadmium. Although these metals are well-known poisonous agents requiring careful handling, their pesticidal capabilities render them useful for given applications, especially in controlled environments [73].

5.1.4 Biosynthesis or bioprospecting

Biosynthesis or bioprospecting of metallic nanoparticles is a new approach that is gaining interest in today's world. Among them, silver and zinc oxide nanoparticles can be created from plant extracts or microbial sources. These techniques help to create effective pesticides and also eliminate the toxicity associated with metal extraction [74].

5.2 Evaluating metals for pesticidal use in ensuring their efficiency, safety, and sustainability

The evaluation of metals for pesticidal use is a critical step in ensuring their effectiveness, safety, and sustainability. This involves several phases:

5.2.1 Laboratory testing

In initial testing, one seeks to ascertain the pesticidal activity of the metal against the targeted pests. Parameters, such as lethal concentration (LC50) and effective concentration (EC50), are calculated. These tests also explore the mode of action of the metal, such as interfering with their reproductive systems or disrupting enzymatic pathways in pests [75].

5.2.2 Field trials

After laboratory testing, the performance of metals is evaluated under actual agricultural conditions. Field trials determine persistence of the metal in soil and plants, effectiveness in different climates, and interactions with other agricultural inputs. For instance, copper-based fungicides are tested for their ability to control fungal diseases like downy mildew under varying humidity and temperature conditions [76].

5.2.3 Environmental impact assessment

These metals should be assessed critically for their possible effects on the environment. The assessment should cover bioaccumulation risks, leaching into groundwater, and toxicity to non-target organisms, such as pollinators, aquatic species, and beneficial soil microbes [75, 76].

5.2.4 Formulation and stability testing

Metal-based pesticides are formulated for improved effectiveness and stability. It includes the blending of metals with inert ingredients, which improve the dispersibility, resistance to sunlight degradation, and adherence to the plant surface. Stability testing is carried out in order to ensure that the product remains effective during different storage and environmental conditions [75, 76].

6. Challenges and risks of using metals for insect control

The use of metals for insect control presents several challenges and risks. Metal toxicity and environmental concerns arise as these metals can accumulate in soil and water, posing threats to non-target organisms like earthworms, beneficial insects, and aquatic species. This can disrupt ecosystems and biodiversity, particularly in areas with high or repeated metal use [77, 78]. Additionally, insects may develop resistance to metal-based treatments, much like they do with chemical pesticides. Over time, pests may evolve mechanisms to tolerate metals, reducing the effectiveness of these strategies [79]. Another challenge is the lack of clear sustainability and regulation surrounding metal nanoparticle use. Regulations as yet to evolve, but there are still inconsistent guidelines, allowing for slow adoption and misuse. Moreso, metals can, inadvertently, harm beneficial species, including pollinators and natural predators, upsetting natural pest control and biodiversity. Metal accumulation can also affect soil and water health resulting in contaminating aquatic ecosystems, and impeding microbial activity required for nutrient cycling [80].

Bioaccumulation of metals, including in nanoparticle form, can also occur and presents long-term health risks. The high cost of production and precise application further limits the widespread use of metal-based treatments, especially in small-scale farmer production. Ultimately, regulatory delays, public perception, and a fear of the safety and environmental impact of such technologies may hold them back from being adopted [77, 80]. These challenges demonstrate the difficulty associated with using metals for the control of pests. They represent promising alternatives to traditional pesticides and should be used sparingly to minimize environmental impact, delay resistance development, and improve sustainable, safe application to the environment.

7. Conclusion

Insects depend on metals for proper survival and physiological functionality, as these metals affect development, reproduction, and interaction with the environment. This chapter highlights the dual roles of metals in pest management: Misregulated as disruptors of insect homeostasis and as critical nutrients for insect metabolism. We also outlined the mechanisms by which metals elicit pesticidal effects, the benefits, and drawbacks of utilizing metals within industries to control pests, and present recent developments in the transition to environmentally benign, metal-based pesticides. Knowing metal ion transporters, metalloprotein enzymes, and mechanisms of detoxification enables researchers to utilize metal homeostasis for developing novel pest-control strategies. Metal-based pesticides, nanoparticles, and chelators are examples of metals used in IPM. In addition, advancements in insect siRNA and biotechnologies, including CRISPR, are poised to enable increasingly directive interventions in metal-regulation pathways of pest insects. While such practices must be prioritized to mitigate risks of environmental toxicity, resistance development, and impacts on non-target species, the benefits of drought-tolerant or limited water use crops shall always outweigh their risks. Research into understanding the complexities of metal's interaction with insects should lead to the development of eco-friendly pesticides. Integration of metals with biological control and other IPM strategies allows us to decrease dependence on chemical insecticides and diversify sustainable agriculture.

Acknowledgements

This work was supported by the Class B Breeding Program of Special Projects for Leading Science and Technology of the Chinese Academy of Sciences (XDPB16), the Fundamental Research Funds for the Central Universities, Beijing Municipal Natural Science Foundation (7202129), the National Natural Science Foundation of China (31571042), the Key Basic Research Project of Applied Basic Research Program of Hebei Province (18966315D), and Hundred Outstanding Creative Talents Support Program of Hebei (BR2-218).

Conflict of interest

The authors declare no conflict of interest.

Author details

Inam Ullah, Hafza Wajeeha Ijaz, Yudie Ma and Minglin Lang*
CAS Center for Excellence in Biotic Interactions, College of Life Science, University of Chinese Academy of Sciences, Beijing, China

*Address all correspondence to: langml@ucas.ac.cn

References

[1] Sawicka B, Egbuna C. Pests of Agricultural Crops and Control Measures. Natural Remedies for Pest, Disease and Weed Control. Amsterdam, Netherlands: Elsevier; 2020. pp. 1-16. DOI: 10.1016/B978-0-12-819304-4.00001-4

[2] Fried G, Chauvel B, Reynaud P, Sache I. Decreases in crop production by non-native weeds, pests, and pathogens. In: Impact of Biological Invasions on Ecosystem Services. Berlin, Germany: Springer; 2017. pp. 83-101. DOI: 10.1007/978-3-319-45121-3_6

[3] Karami-Mohajeri S, Abdollahi M. Toxic influence of organophosphate, carbamate, and organochlorine pesticides on cellular metabolism of lipids, proteins, and carbohydrates: A systematic review. Human & Experimental Toxicology. 2011;**30**(9):1119-1140. DOI: 10.1177/0960327110388959

[4] López-Pantoja G, Paramio AM, Malia S, Sánchez-Osorio I. Candidate natural enemy assemblage for the biological control of *Lymantria dispar* L. in oak (*Quercus* spp.) with different levels of pest infestation. Journal of Forest Science. 2024;**70**(8):420-435. DOI: 10.17221/14/2024-JFS

[5] Bueno VH, Montes FC, Costaz T, De Jong PW, Parra JR, Van Lenteren JC. Do releases of the mirid predator Macrolophus basicornis (Hemiptera: Miridae) together with the egg parasitoid Trichogramma pretiosum (hymenoptera: Trichogrammatidae) improve biological control of Tuta absoluta (Lepidoptera: Gelechiidae) in tomato? Journal of Economic Entomology. 2023;**116**(3):733-741. DOI: 10.1093/jee/toad060

[6] Khayal AY, Ullah I, Nughman M, Shah SM, Ali N. Elemental, antimicrobial and antioxidant activities of a medicinal plant Sorghum halepense. Pure and Applied Biology. 2019;**8**(1):795-803. DOI: 10.19045/bspab.2019.80022

[7] Reddy PP. Agro-Ecological Approaches to Pest Management for Sustainable Agriculture. Berlin, Germany: Springer; 2017. DOI: 10.1007/978-981-10-4325-3

[8] Ozalp O, Uzcan F, Gumus ZP, Soylak M. Sample preparation methods for metal containing pesticides in food and environmental samples. Critical Reviews in Analytical Chemistry. 2024;**54**(5):1109-1120. DOI: 10.1080/10408347.2022.2106118

[9] Khan S, Lang M. A comprehensive review on the roles of metals mediating insect–microbial pathogen interactions. Metabolites. 2023;**13**(7):839. DOI: 10.3390/metabo13070839

[10] Singh A, Dhiman N, Kar AK, Singh D, Purohit MP, Ghosh D, et al. Advances in controlled release pesticide formulations: Prospects to safer integrated pest management and sustainable agriculture. Journal of Hazardous Materials. 2020;**385**:121525. DOI: 10.1016/j.jhazmat.2019.121525

[11] Jensen P, Trumble JT. Ecological consequences of bioavailability of metals and metalloids in insects. Recent Research Development Entomology. 2003;**42**:1-17. DOI: 10.5555/20043113648

[12] Gao Q, Wang X, Wang W, Lei C, Zhu F. Influences of chromium and cadmium on the development of black soldier fly larvae. Environmental Science and Pollution Research. 2017;**24**(9):8637-8644. DOI: 10.1007/s11356-017-8550-3

[13] Huibo Z, Yong Z, Rui L, Guorui L, Jianjun D, Qi W, et al. Analysis of the mechanism of *Ricinus communis* L. tolerance to Cd metal based on proteomics and metabolomics. PLoS One. 2023;**18**(3):e0272750. DOI: 10.1371/journal.pone.0272750

[14] De Almeida A, Oliveira BL, Correia JD, Soveral G, Casini A. Emerging protein targets for metal-based pharmaceutical agents: An update. Coordination Chemistry Reviews. 2013;**257**(19-20):2689-2704. DOI: 10.1016/j.ccr.2013.01.031

[15] Su H, Wu J, Zhang Z, Ye Z, Chen Y, Yang Y. Effects of cadmium stress at different concentrations on the reproductive behaviors of beet armyworm Spodoptera exigua (Hübner). Ecotoxicology. 2021;**30**:402-410. DOI: 10.1007/s10646-021-02365-x

[16] Dow JA. The essential roles of metal ions in insect homeostasis and physiology. Current Opinion in Insect Science. 2017;**23**:43-50. DOI: 10.1016/j.cois.2017.07.001

[17] Beeby A. Toxic metal uptake and essential metal regulation in terrestrial invertebrates: A review. In: Metal Ecotoxicology Concepts and Applications. United Kingdom: Taylor & Francis; 2020. pp. 65-89. DOI: 10.1201/9781003069973-3

[18] Van der Fels-Klerx H, Camenzuli L, Van Der Lee M, Oonincx D. Uptake of cadmium, lead and arsenic by Tenebrio molitor and Hermetia illucens from contaminated substrates. PLoS One. 2016;**11**(11):e0166186. DOI: 10.1371/journal.pone.0166186

[19] Maya-Maldonado K, Cardoso-Jaime V, González-Olvera G, Osorio B, Recio-Tótoro B, Manrique-Saide P, et al. Mosquito metallomics reveal copper and iron as critical factors for plasmodium infection. PLoS Neglected Tropical Diseases. 2021;**15**(6):e0009509. DOI: 10.1371/journal.pntd.0009509

[20] Balistrieri LS, Mebane CA, Schmidt TS. Time-dependent accumulation of Cd, Co, Cu, Ni, and Zn in natural communities of mayfly and caddisfly larvae: Metal sensitivity, uptake pathways, and mixture toxicity. Science of the Total Environment. 2020;**732**:139011. DOI: 10.1016/j.scitotenv.2020.139011

[21] Liu Y, Zhu F, Shen Z, Moural TW, Liu L, Li Z, et al. Glutaredoxins and thioredoxin peroxidase involved in defense of emamectin benzoate induced oxidative stress in Grapholita molesta. Pesticide Biochemistry and Physiology. 2021;**176**:104881. DOI: 10.1016/j.pestbp.2021.104881

[22] Liu Y, Du C, Lin C, Gao X, Zhu J, Zhang C. Characterization of copper/zinc superoxide dismutase activity on Phascolosoma esculenta (Sipuncula: Phascolosomatidea) and its protection from oxidative stress induced by cadmium. International Journal of Molecular Sciences. 2022;**23**(20):12136. DOI: 10.3390/ijms232012136

[23] Cui S, Wang L, Ma L, Geng X. P450-mediated detoxification of botanicals in insects. Phytoparasitica. 2016;**44**:585-599. DOI: 10.1007/s12600-016-0550-1

[24] Garza-Lombó C, Posadas Y, Quintanar L, Gonsebatt ME, Franco R. Neurotoxicity linked to dysfunctional metal ion homeostasis and xenobiotic metal exposure: Redox signaling and oxidative stress. Antioxidants & Redox Signaling. 2018;**28**(18):1669-1703. DOI: 10.1089/ars.2017.7272

[25] Hrdina A, Iatsenko I. The roles of metals in insect–microbe interactions and immunity. Current Opinion in Insect

Science. 2022;**49**:71-77. DOI: 10.1016/j.cois.2021.12.004

[26] Ullah I, Lang M. Key players in the regulation of iron homeostasis at the host-pathogen interface. Frontiers in Immunology. 2023;**14**:1279826. DOI: 10.3389/fimmu.2023.1279826

[27] Ali S, Ahmad N, Dar MA, Manan S, Rani A, Alghanem SMS, et al. Nano-agrochemicals as substitutes for pesticides: Prospects and risks. Plants. 2023;**13**(1):109. DOI: 10.3390/plants13010109

[28] Ali S, Ullah MI, Sajjad A, Shakeel Q, Hussain A. Environmental and health effects of pesticide residues. Sustainable agriculture reviews 48: Pesticide occurrence, analysis and remediation Vol 2. Analysis. 2021;**2**:311-336. DOI: 10.1007/978-3-030-54719-6_8

[29] NourEldin BM, Gahlan AA, Mahross MH, Abdelhameed RM. The behavior of mixed metal based copper–organic framework for uptake of chlorpyrifos pesticide from wastewater and its antimicrobial activity. Water, Air, & Soil Pollution. 2024;**235**(11):746. DOI: 10.1007/s11270-024-07515-5

[30] Yan S, Tan M, Zhang A, Jiang D. The exposure risk of heavy metals to insect pests and their impact on pests occurrence and cross-tolerance to insecticides: A review. Science of the Total Environment. 2024;**916**:170274. DOI: 10.1016/j.scitotenv.2024.170274

[31] Singh N, Gupta VK, Kumar A, Sharma B. Synergistic effects of heavy metals and pesticides in living systems. Frontiers in Chemistry. 2017;**5**:70. DOI: 10.3389/fchem.2017.00070

[32] Jiang D, Yan S. Effects of Cd, Zn or Pb stress in Populus alba berolinensis on the development and reproduction of Lymantria dispar. Ecotoxicology. 2017;**26**(10):1305-1313. DOI: 10.1007/s10646-017-1855-7

[33] Burandt QC, Deising HB, von Tiedemann A. Further limitations of synthetic fungicide use and expansion of organic agriculture in Europe will increase the environmental and health risks of chemical crop protection caused by copper-containing fungicides. Environmental Toxicology and Chemistry. 2024;**43**(1):19-30. DOI: 10.1002/etc.5766

[34] Seku K, Hussaini SS, Reddy GB, Reddy MRK. Silver-based biofungicides for the suppression of pathogenic fungi in agriculture fields. In: Nanofungicides. Amsterdam, Netherlands: Elsevier; 2024. pp. 169-194. DOI: 10.1016/B978-0-323-95305-4.00009-1

[35] Eason C, Ross J, Blackie H, Fairweather A. Toxicology and ecotoxicology of zinc phosphide as used for pest control in New Zealand. New Zealand Journal of Ecology. 2013;**37**(1):1-11. Available from: https://www.jstor.org/stable/24060751

[36] Bernard CE, Harrass MC, Manning MJ. Boric acid and inorganic borate pesticides. In: Hayes' Handbook of Pesticide Toxicology. Amsterdam, Netherlands: Elsevier; 2010. pp. 2033-2053. DOI: 10.1016/B978-0-12-374367-1.00094-X

[37] Mason CJ, Grummer A, Bosch M, Shikano I. Adult dietary experience influences mortality of the pest melon fly, Zeugodacus cucurbitae (Diptera: Tephritidae), to an ingested toxin. Physiological Entomology. 2024;**50**(1):77-87. DOI: 10.1111/phen.12468

[38] Rae RG, Robertson JF, Wilson MJ. Optimization of biological

(Phasmarhabditis hermaphrodita) and chemical (iron phosphate and metaldehyde) slug control. Crop Protection. 2009;**28**(9):765-773. DOI: 10.1016/j.cropro.2009.04.005

[39] Moghadamnia AA. An update on toxicology of aluminum phosphide. DARU Journal of Pharmaceutical Sciences. 2012;**20**:1-8. DOI: 10.1186/2008-2231-20-25

[40] Stuart AM, Jacob J, Awoniyi AM, Costa F, Bosma L, Meheretu Y, et al. Alternative domestic rodent pest management approaches to address the hazardous use of metal phosphides in low- and middle-income countries. Journal of Pest Science. 2024;**98**:1-23. DOI: 10.1007/s10340-024-01825-7

[41] Rawtani D, Khatri N, Tyagi S, Pandey G. Nanotechnology-based recent approaches for sensing and remediation of pesticides. Journal of Environmental Management. 2018;**206**:749-762. DOI: 10.1016/j.jenvman.2017.11.037

[42] Rajesh A, Madhumitha G. An insight into the insecticidal activity of green synthesized silver nanoparticles. Colloid Journal. 2023;**85**(5):854-870. DOI: 10.1134/S1061933X23600045

[43] Rai M, Ingle AP, Pandit R, Paralikar P, Shende S, Gupta I, et al. Copper and copper nanoparticles: Role in management of insect-pests and pathogenic microbes. Nanotechnology Reviews. 2018;**7**(4):303-315. DOI: 10.1515/ntrev-2018-0031

[44] Ramkumar G, Asokan R, Ramya S, Gayathri G. Characterization of Trigonella foenum-graecum derived iron nanoparticles and its potential pesticidal activity against Tuta absoluta (Lepidoptera). Journal of Cluster Science. 2021;**32**:1185-1190. DOI: 10.1007/s10876-020-01867-8

[45] Kumaravel J, Lalitha K, Arunthirumeni M, Shivakumar MS. Mycosynthesis of bimetallic zinc oxide and titanium dioxide nanoparticles for control of Spodoptera frugiperda. Pesticide Biochemistry and Physiology. 2021;**178**:104910. DOI: 10.1016/j.pestbp.2021.104910

[46] Baker BP, Green TA, Loker AJ. Biological control and integrated pest management in organic and conventional systems. Biological Control. 2020;**140**:104095. DOI: 10.1016/j.biocontrol.2019.104095

[47] Galli M, Feldmann F, Vogler UK, Kogel K-H. Can biocontrol be the game-changer in integrated pest management? A review of definitions, methods and strategies. Journal of Plant Diseases and Protection. 2024;**131**(2):265-291. DOI: 10.1007/s41348-024-00878-1

[48] Peng Y, Li SJ, Yan J, Tang Y, Cheng JP, Gao AJ, et al. Research progress on phytopathogenic fungi and their role as biocontrol agents. Frontiers in Microbiology. 2021;**12**:670135. DOI: 10.3389/fmicb.2021.670135

[49] Li C, Li Y, Ding C. The role of copper homeostasis at the host-pathogen axis: From bacteria to fungi. International Journal of Molecular Sciences. 2019;**20**(1):175. DOI: 10.3390/ijms20010175

[50] Ullah I, Ali N, Ullah W, Qasim M, Nughman M, Ullah N. Partial purification, characterization and application of thermoalkaliphilic proteases from Priestia endophytica, Lysinibacillus cresolivorans and Bacillus subtilis isolated from desert soil. Malaysian. Journal of Microbiology. 2022;**18**(3):251-260. DOI: 10.21161/mjm.211323

[51] Gardiner MM, Harwood JD. Influence of heavy metal contamination

on urban natural enemies and biological control. Current Opinion in Insect Science. 2017;**20**:45-53. DOI: 10.1016/j.cois.2017.03.007

[52] Lin Y, Xu J, Zhang Y. Identification method of citrus aurantium diseases and pests based on deep convolutional neural network. Computational Intelligence and Neuroscience. 2022;**2022**(1):7012399. DOI: 10.1155/2022/7012399

[53] Adhikari U. Insect pest management: Mechanical and physical techniques. Reviews in Food and Agriculture (RFNA). 2022;**3**(1):48-53. DOI: 10.26480/rfna.01.2022.48.53

[54] Bell C. Pest control of stored food products: Insects and mites. In: Hygiene in Food Processing. Amsterdam, Netherlands: Elsevier; 2014. pp. 494-538. DOI: 10.1533/9780857098634.3.494

[55] Mirjalili S, Mirjalili SM, Lewis A. Grey wolf optimizer. Advances in Engineering Software. 2014;**69**:46-61. DOI: 10.1016/j.advengsoft.2013.12.007

[56] Altieri MA, Nicholls CI, Montalba R. Technological approaches to sustainable agriculture at a crossroads: An agroecological perspective. Sustainability. 2017;**9**(3):349. DOI: 10.3390/su9030349

[57] Josephrajkumar A, Mani M, Anes K, Mohan C. Ecological engineering in pest management in horticultural and agricultural crops. Trends in Horticultural Entomology. 2022:123-155. DOI: 10.1007/978-981-19-0343-4_4

[58] Cantó-Tejero M, Casas JL, Marcos-García MÁ, Pascual-Villalobos MJ, Florencio-Ortiz V, Guirao P. Essential oils-based repellents for the management of Myzus persicae and Macrosiphum euphorbiae. Journal of Pest Science. 2022;**95**(1):365-379. DOI: 10.1007/s10340-021-01380-5

[59] Beizhou S, Jie Z, Jinghui H, Hongying W, Yun K, Yuncong Y. Temporal dynamics of the arthropod community in pear orchards intercropped with aromatic plants. Pest Management Science. 2011;**67**(9):1107-1114. DOI: 10.1002/ps.2156

[60] Kumar A, Prakash A, Johri B. Bacillus as PGPR in crop ecosystem. In: Bacteria in Agrobiology: Crop Ecosystems. Berlin, Germany: Springer; 2011. pp. 37-59. DOI: 10.1007/978-3-642-18357-7_2

[61] Veliz EA, Martínez-Hidalgo P, Hirsch AM. Chitinase-producing bacteria and their role in biocontrol. AIMS Microbiology. 2017;**3**(3):689. DOI: 10.3934/microbiol.2017.3.689

[62] Kilani-Morakchi S, Morakchi-Goudjil H, Sifi K. Azadirachtin-based insecticide: Overview, risk assessments, and future directions. Frontiers in Agronomy. 2021;**3**:676208. DOI: 10.3389/fagro.2021.676208

[63] Chaudhari AK, Singh VK, Kedia A, Das S, Dubey NK. Essential oils and their bioactive compounds as eco-friendly novel green pesticides for management of storage insect pests: Prospects and retrospects. Environmental Science and Pollution Research. 2021;**28**:18918-18940. DOI: 10.1007/s11356-021-12841-w

[64] Yarzábal LA, Chica EJ. Role of rhizobacterial secondary metabolites in crop protection against agricultural pests and diseases. In: New and Future Developments in Microbial Biotechnology and Bioengineering. Amsterdam, Netherlands: Elsevier; 2019. pp. 31-53. DOI: 10.1016/B978-0-444-63504-4.00003-7

[65] Xu X, Harvey-Samuel T, Siddiqui HA, Ang JXD, Anderson ME, Reitmayer CM, et al. Toward a CRISPR-Cas9-based gene drive in the diamondback moth Plutella xylostella. 2022;**5**(2):224-236. DOI: 10.1089/crispr.2021.0129

[66] Hendrichs J, Franz G, Rendon P. Increased effectiveness and applicability of the sterile insect technique through male-only releases for control of Mediterranean fruit flies during fruiting seasons. Journal of Applied Entomology. 1995;**119**(1-5):371-377. DOI: 10.1111/j.1439-0418.1995.tb01303.x

[67] Gao Y, Alyokhin A, Prager SM, Reitz S, Huseth A. Complexities in the implementation and maintenance of integrated pest management in potato. Annual Review of Entomology. 2024;**70**:45-63. DOI: 10.1146/annurev-ento-120523-023156

[68] Christiaens O, Sweet J, Dzhambazova T, Urru I, Smagghe G, Kostov K, et al. Implementation of RNAi-based arthropod pest control: Environmental risks, potential for resistance and regulatory considerations. Journal of Pest Science. 2022;**95**(1):1-15. DOI: 10.1007/s10340-021-01439-3

[69] Agrawal J, Arafat MY. Transforming farming: A review of AI-powered UAV technologies in precision agriculture. Drones. 2024;**8**(11):664. DOI: 3390/drones8110664

[70] Qazi S, Khawaja BA, Farooq QU. IoT-equipped and AI-enabled next generation smart agriculture: A critical review, current challenges and future trends. IEEE Access. 2022;**10**:21219-21235. DOI: 10.1109/ACCESS.2022.3152544

[71] Mesquita AF, Gonçalves FJ, Gonçalves AM. The lethal and sub-lethal effects of fluorinated and copper-based pesticides—A review. International Journal of Environmental Research and Public Health. 2023;**20**(4):3706. DOI: 10.3390/ijerph20043706

[72] Matinde E, Simate GS, Ndlovu S. Mining and metallurgical wastes: A review of recycling and re-use practices. Journal of the Southern African Institute of Mining and Metallurgy. 2018;**118**(8):825-844. DOI: 10.17159/2411-9717/2018/v118n8a5

[73] Oladimeji TE, Oyedemi M, Emetere ME, Agboola O, Adeoye JB, Odunlami OA. Review on the impact of heavy metals from industrial wastewater effluent and removal technologies. Heliyon. 2024;**10**(23). DOI: 10.1016/j.heliyon.2024.e40370

[74] Jeevanandam J, Kiew SF, Boakye-Ansah S, Lau SY, Barhoum A, Danquah MK, et al. Green approaches for the synthesis of metal and metal oxide nanoparticles using microbial and plant extracts. Nanoscale. 2022;**14**(7):2534-2571. DOI: 10.1039/D1NR08144F

[75] Paunescu A, Ponepal CM, Tofan L, Brinzea G, Tantu MM, Mihaescu CF, et al. Ecotoxicological risk assessment of Actellic 50 EC insecticide on non-target organisms in parallel with the application of standardized tests. Toxics. 2022;**10**(12):745. DOI: 10.3390/toxics10120745

[76] Chapman PM, Wang F, Janssen CR, Goulet RR, Kamunde CN. Conducting ecological risk assessments of inorganic metals and metalloids: Current status. Human and Ecological Risk Assessment. 2003;**9**(4):641-697. DOI: 10.1080/713610004

[77] Alengebawy A, Abdelkhalek ST, Qureshi SR, Wang M-Q. Heavy metals and pesticides toxicity in agricultural soil and plants: Ecological risks and human

health implications. Toxics. 2021;**9**(3):42. DOI: 10.3390/toxics9030042

[78] Wallace DR, Djordjevic AB. Heavy metal and pesticide exposure: A mixture of potential toxicity and carcinogenicity. Current Opinion in Toxicology. 2020;**19**:72-79. DOI: 10.1016/j.cotox.2020.01.001

[79] Kráľová K, Jampílek J. Metal- and metalloid-based nanofertilizers and nanopesticides for advanced agriculture. In: Inorganic Nanopesticides and Nanofertilizers: A View from the Mechanisms of Action to Field Applications. Berlin, Germany: Springer; 2022. pp. 295-361. DOI: 10.1007/978-3-030-94155-0_10

[80] Yamini V, Shanmugam V, Rameshpathy M, Venkatraman G, Ramanathan G, Garalleh HA, et al. Environmental effects and interaction of nanoparticles on beneficial soil and aquatic microorganisms. Environmental Research. 2023;**236**:116776. DOI: 10.1016/j.envres.2023.116776